Fire in the Jungle

Fire in the Jungle:
A Study of One of America's Most Successful Unconventional Warfare Campaigns

Colonel (Ret.) Larry S. Schmidt

Edited by Paul D. LeFavor

Blacksmith Publishing

Fayetteville, North Carolina

"The people who have not yet been conquered by the enemy will be the most eager to arm against him; they will set an example that will gradually be followed by their neighbors. **The flames will spread like a brush fire**, until they reach the area on which the enemy is based, threatening his lines of communication and his very existence." – Clausewitz

Fire in the Jungle: A Study of One of America's Most
Successful Unconventional Warfare Campaigns
by Colonel (Ret.) Larry S. Schmidt
edited by Paul D. LeFavor

Copyright © 2018 Larry S. Schmidt

ISBN 978-0-9977434-5-6

Library of Congress Control Number: 2018914967

Printed in the United States of America

Published by Blacksmith LLC
Fayetteville, North Carolina

www.BlacksmithPublishing.com

Direct inquiries and/or orders to the above web address.

Contents

Foreword viii

Preface x

Prolegomena xiv

Chronology xix

The Seven Phases of the Mindanao Resistance xii

Glossary of Terms xxiii

Dramatis Personae xxv

Maps xxvii

Introduction 1

Chapter

1 – The Commonwealth of the Philippines 8

2 – Run to the Hills! 16

3 – Co-Prosperity Sphere 30

4 – "General" Fertig 38

5 – The Krag and the Kris 52

6 – MacArthur's Unconventional War 64

7 – Fertig's Fiefdom 70

8 – 10th Military District 78

9 – The Empire Strikes Back 94

10 – Kempeitai, Kalibapi and the Underground 104

11 – Beans, Bullets and Submarines 115

12 – Cargadores, Coconut Oil and Cannons 134

13 – War of the Flea 145

14 – The Jungle Telegraph 156

15 – Victor V 162

16 – Conclusion 175

Appendix A: 10th Military Division Units................... 186

Bibliography... 190

Index 194

About the Author.. 197

To the American Guerrillas of Mindanao with whom I have corresponded and spoken. May this brief history recognize in some small manner the duty and patriotism your courageous actions have so demonstrated.

Foreword

"There is another type of warfare—new in its intensity, ancient in its origin—war by guerrillas, subversives, insurgents, assassins; war by ambush instead of by combat, by infiltration instead of aggression, seeking victory by eroding and exhausting the enemy instead of engaging him. It preys on unrest." – President John F. Kennedy, 1962

I have been privileged to serve and defend this nation among the best, elite, and unconventional forces our nation has in its arsenal. From the 75th Ranger Regiment (Airborne) to the Special Forces (SF) – Green Berets. Of my 33-year career in the US Army, 27 years were spent serving as a Green Beret, living by, with, and through foreign indigenous populations and executing all seven phases of the Unconventional Warfare (UW) framework across the SF operational continuum. It took me years to understand the history, nature, and application of UW, culminating in my last two years as the UW Manager for the United States Special Operations Command (USSOCOM).

Throughout history, the resistance of a conquered people against the foreign invader has become one of the most romanticized of events. The term "resistance movement," according to Colonel Russell Volckmann, may be used to describe collectively "the discontented elements of a populace who by various methods oppose and operate against established civil and military authority." Many resistance movements have become the inspiration for national folklore and have been celebrated through the writings of novelists and historians. The Tyrolese uprising against Napoleon's armies, the legendary French Maquis, and the Yugoslav "Sons of the Eagle" are well-known examples. From these accounts, a researcher can derive a paradigm for a successful resistance movement. While, Americans are familiar with the resistance movements in Europe during WWII, specifically those led by the Office of Strategic Services (OSS), few are aware of the resistance of the Filipinos against the Japanese. Notably, resistance operations in the Philippines were of paramount importance to the formation of US Army Special Forces and its originating doctrine. For example, Colonel Russell Volckmann, a contemporary of Colonel Wendell Fertig, wrote FM 31-20, *Operations Against Guerrilla Forces* and FM 31-21, *Organization and Conduct of Guerrilla Warfare.*

As noted, the resistance movement in the Philippines against the Japanese is largely unsung and unheard. This book was written to bring that awareness and demonstrate how the resistance against the Japanese, particularly on the Island of Mindanao, was prototypic of a successful resistance movement. Larry S. Schmidt and Paul D. LeFavor make a significant contribution to the study of resistance by telling this story. It therefore warrants my strongest recommendation to the reader. Study the past, and you will define the future.

De Oppresso Liber
Chief Warrant Officer Five Charles Moritz, USA (Ret.)

Preface

"The guerrilla movement has become one of the greatest romantic themes of subsequent Philippine history and lore because for the Filipinos, the resistance was one of the finest hours for the Philippine people." – David Steinberg

After a four-month struggle, American and Filipino forces in the Philippines capitulated to the Imperial Japanese forces. After "liberating" the Philippines, Japan incorporated the island archipelago into the Greater East Asia Co-Prosperity Sphere. As the occupation commenced, a resistance movement was born among the defeated people.

The Philippine resistance was in fact "the finest hour" for many Americans who did not surrender but rather took to the hills, joining Filipino guerrillas to fight the Japanese. As with many guerrilla resistance groups elsewhere in World War II, many of these units were led by Americans. The Philippine resistance movement, however, gives us the first historical example of Americans, military and civilian, organizing guerrilla units on a grand scale. The movement is also prototypical of an effective, coordinated guerrilla resistance, and it is well worth studying for that reason, if for no other.

There were many guerrilla organizations operating throughout the 1,000-mile-long Philippine Archipelago. They had names like "Blackburn's Headhunters," "Marking's Guerrillas," "President Quezon's Own Guerrillas," "Lawin's Patriot and Suicide Forces," and "The Live or Die Unit," among the many. Some of these groups had an almost opera bouffe character, and others were complete and formal organizations, down to training camps, maneuvers, CCS, orders of battle, and the usual military red tape. In all, some 260,715 guerrillas in 277 guerrilla units fought in the resistance movement as organized, armed, and tactically employed units.

Among the many guerrilla units in the Philippines was the 10th Military District, the Mindanao guerrillas, commanded by Colonel Wendell W. Fertig. Fertig had been a United States Army Reserve officer and mining engineer before the war and had found himself in a position to lead the guerrilla resistance on the island of Mindanao in the Southern Philippines.

This book studies the Mindanao guerrilla organization for several reasons. The subordinate organizations of Fertig's 10th Military District were predominantly American led. This fact makes its study worthwhile not only because it was unique among the guerrilla organizations in the Philippines but also because it provides a good historical example to measure concepts for the support of resistance movements. Furthermore, the Mindanao guerrilla organization is generally illustrative of the growth of the other guerrilla organizations throughout the Philippine Islands, so a study of it provides clues as to how the other guerrilla organizations were established and sustained.

Americans - military, foreign service, or civilian - may find themselves isolated unexpectedly in territory occupied by a nation at war with either the United States or a nation allied with the United States. Initially, the decision which must be made by such an individual is whether or not to surrender to the enemy, and there are many factors which will influence that decision. If the individual chooses not to surrender, then as a person non-indigenous to the occupied population, entirely unforeseen hardships may be encountered.

Cunning and ingenuity coupled with strong courage and loyalty are the personal qualities which will become mandatory for survival. Beyond these personal traits, an understanding of how resistance movements grow and achieve success will be highly useful. The decision to resist is, for the people of a conquered nation, a political decision and therefore has ramifications much beyond the basic decision to survive. This knowledge will also be useful to American military and State Department planners as they measure the potential of a resistance movement to contribute to the overall strategic and tactical plans of the regular forces.

The Philippine resistance movement has not been studied in any detail by American historians, yet the United States played a vital role in the organization and ultimate success of the movement. This book seeks to fill some of this void by contributing a historical study of one organization within the resistance movement - the 10th Military District on Mindanao.

On December 28, 1941, President Franklin D. Roosevelt broadcast the following message to the people of the Philippines:

> I give the people of the Philippines my solemn pledge that their freedom will be redeemed and their independence established and protected. The entire resources, in men and material, of the United States stand behind that pledge.[1]

It was two years and ten months later before that pledge was fulfilled. In the meantime, guerrillas - Filipinos and Americans together – fought a desperate

---

[1] Catherine Porter, *Crisis in the Philippines* (New York: A. A. Knopf, 1942), 104.

resistance against a cruel conqueror against very long odds of achieving success. In General Douglas MacArthur's tribute to the Filipino guerrillas, he said:

> We are aided by the militant loyalty of a whole people – a people who have rallied as one behind the standards of those stalwart patriots who, reduced to wretched material conditions yet sustained by an unconquerable spirit, hate formed an invincible center to a resolute overall resistance.[2]

This book is the story of that "invincible center."

Major Larry S. Schmidt
Fort Leavenworth, KS
Spring 1982

---

[2] General Headquarters, Southwest Pacific Area, "Special Release," October 25, 1944.

Prolegomena

"The combination of guerrilla and subversive war has been pursued with increasing success in the neighboring areas of South-East Asia and in other parts of the world. Campaigns of this kind are likely to continue because they fit the conditions of the modern age and at the same time are well suited to take advantage of social discontent, racial ferment, and nationalistic fervor." – B.H. Liddell Hart

This book is about guerrilla warfare, which as a subset of unconventional warfare, is the overt military aspect of a resistance or insurgency. Current US doctrine defines unconventional warfare (UW) as "activities conducted to enable a resistance movement or insurgency to coerce, disrupt, or overthrow a government or occupying power by operating through or with an underground, auxiliary, and guerrilla force in a denied area."[3] The counterpart to UW is what is known in US doctrine as foreign internal defense (FID). An aspect of FID is counterinsurgency (COIN). This relationship is seen in the diagram below.

Unconventional Warfare	Foreign Internal Defense
Coerce, disrupt, or overthrow a government or occupying power through: policy, indigenous force, or subversion/sabotage.	Improve a nation-state's security apparatus by: training, advising, and assisting primarily through COIN.
Degrade legitimacy and destabilize.	Reinforce legitimacy and stabilize.

---

[3] TC 18-01, *Unconventional Warfare* (Washington, D.C.: Government Printing Office, 2011), 1-1.

Further, a resistance movement may be divided into three comparatively distinct elements (see figure 1). Although the categories tend to blur as one moves along the continuum, the three elements are nevertheless discernible and the definitions workable. The three elements are the guerrilla force, the auxiliary, and the underground.[4]

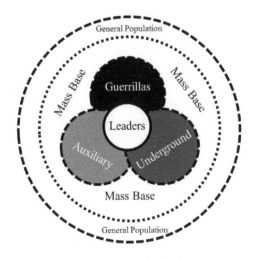

Figure 1. Three Components of a Resistance Movement.

As postulated by Che Guevara, "The guerrilla band is an armed nucleus, the fighting vanguard of the people."[5] As the vanguard of the people, the guerrilla force conducts military and paramilitary operations by irregular, predominately indigenous forces in enemy-held or hostile territory. It is the overt military aspect of

---

[4] TC 18-01, 1-1.

[5] Che Guevara, *Guerrilla Warfare* (Lincoln, NE: University of Nebraska Press, 1985), 52.

an insurgency or armed resistance. It is often rurally based, with full-time and part-time membership.

The underground supports the area command, auxiliary, and guerrilla force. As the internal support element of the resistance movement, the underground is clandestine in nature, that is, its members do not openly indicate their sympathy or involvement with the resistance movement. While the mission of the underground is sabotage, gathering intelligence and committing acts of deception, the primary mission of the auxiliary is to provide logistical support, security, and intelligence for the guerrilla force by organizing civilian supporters of the resistance movement.

The auxiliary is the internal support element of the resistance movement whose organization and operation are both covert and clandestine in nature and whose members do not openly indicate their sympathy or involvement with the resistance movement.

Seven Phases – The seven-phase unconventional warfare (UW) framework is a conceptual construct that aids in planning.[6] It depicts the normal phases of a UW operation. As the basis for what would be become US doctrine, this is how the phases played out in the Philippines:

A. Latent Phase [7]

   1. Preparation – intelligence preparation, planning, and shaping activities.

---

[6] TC 18-01, *Unconventional Warfare* (Washington, D.C.: Government Printing Office, 2011), 1-9.
[7] Insurgencies normally progress through three phases: latent or incipient, guerrilla warfare, and war of movement (FM 3-24.2 Tactics in Counterinsurgency, 2-15).

a) Began well before the war with Corps of Intelligence Police.
b) Operation Ferdinand and coastal watchers.
c) Allied Intelligence Bureau and the OSS.

2. Initial Contact – coordination is made with resistance or government in exile.

a) Within days of fall of islands, the government was rescued and established in exile.

3. Infiltration – infiltration is made to establish communications with resistance movement.

a) The unsurrendered took to the hills.
b) Parsons' mission.

4. Organization – the resistance is organized, trained, and equipped into three basic components: guerillas, underground, and auxiliary.

a) Structure based on previous existing Military Districts.
b) MacArthur's 'lay-low' order.

B. Guerrilla Warfare

5. Build-up – development of infrastructure: assistance given to cadre with expansion into effective organization.

a) Lasted the longest of all phases.
b) Took the longest time due to the conduct of UW as a shaping operation.

C. *War of Movement*

6. Employment – resistance forces conduct combat
operations.

a) Prior to invasion, mostly functioned as an intelligence
source capable of minor disruption ops.
b) Orders from MacArthur to raise hell.
c) Japanese had grossly underestimated the capabilities
of the Resistance.
d) Progressed very quickly because the Coalition Forces
had arrived and the population was psychologically
prepared to overthrow the Japanese government.

7. Transition

a) Demobilization Plan
b) Governmental compensation to Resistance
Members.[8]

---

[8] I am indebted to Captain Andrew Kroening for this insightful way of
fleshing out these phases.

Chronology

1941

Jul 26	USAFFE formed
Dec 8	Japan attacks the Philippines
Dec 20	Japanese forces land at Davao

1942

Jan 3	HQs, Visayan-Mindanao Force relocated from Cebu to Del Monte, Mindanao
Mar 4	Visayan-Mindanao Force divided into two commands
Mar 13-16	Gen MacArthur on Mindanao. Reviews plans for guerrilla resistance
Apr 9	Bataan falls
Apr 29	Japanese forces land at Cotabato and Parang on Mindanao
Apr 30	Col. Wendell Fertig arrives on Mindanao
May 2-3	Japanese forces land at Cagayan and Bugo on Mindanao
May 6-7	Corregidor falls
May 7	Gen Wainwright orders surrender of all United States and Philippine soldiers in the Philippines
May 10	Gen. Sharp surrenders the Visayan-Mindanao Force
Jun	Captain Luis Morgan begins consolidation of Guerrilla bands in Misamis Oriental Province
Sep 12	Morgan offers command to Fertig
Sep 18	Fertig establishes Mindanao-Sulu Force
Oct 7	Fertig establishes guerrilla HQ in Misamis
Nov	Morgan departs on unification expedition
Dec 4	Captains Smith and Hamner sail to Australia

1943

Jan 1	Misamis Occidental, Zamboanga now secured
Jan 23	105th Division established
Jan 31	KZOM establishes contact with San Francisco
Feb 13	MacArthur appoints Fertig CO of the 10th Military District. Radio communication with GHQ, SWPA
Mar 5	The first GHQ, SWPA intelligence teams and supplies arrive on Mindanao by submarine
Apr 14	10 Americans escape from Davao Penal Colony
Jun	Morgan expedition returns
Jun 26	Japanese attack in force at Misamis. 10th Military District HQ moved to Liangen, Lanao Province
Jul	Morgan revolts, forms Mindanao and Dutch Indies Command
Sep	Morgan deported to Australia
Oct	Laurel government installed, 120-day amnesty period for guerrillas
Oct 7	Salipada Pendatun concedes authority to Fertig
Nov	10th Military District HQ moved to Esperanza, Agusan Province
Dec	Japanese issue proclamation that all unsurrendered Americans will be summarily executed by January 25, 1944.

1944

Jan 1	"A" Corps established. Sulu Archipelago command separated from 10th Military District
Jul	10th Military District HQ moved to Waloe, Agusan Province
Aug-Sep	First bombings of Davao City

1945

Mar-Sep	All guerrilla units deactivated
Apr 18	United States Forces invade Mindanao
Sep 7	Japanese Forces on Mindanao formally surrender
Sep 15	6th Infantry Division (PA) reactivated

The Seven Phases of the Mindanao Resistance

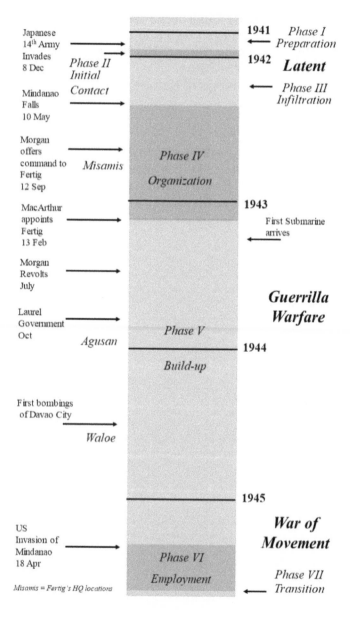

Japanese 14th Army Invades 8 Dec

Phase II Initial Contact

Mindanao Falls 10 May

Morgan offers command to Fertig 12 Sep — Misamis

Phase IV Organization

MacArthur appoints Fertig 13 Feb

Morgan Revolts July

Laurel Government Oct — Agusan

Phase V Build-up

First bombings of Davao City — Waloe

US Invasion of Mindanao 18 Apr

Phase VI Employment

Misamis = Fertig's HQ locations

1941 Phase I Preparation

1942 Latent

Phase III Infiltration

1943 First Submarine arrives

Guerrilla Warfare

1944

1945 War of Movement

Phase VII Transition

Glossary of Terms

Abaca - hemp
Amok - a desperate impulsive frenzied killing; Moro custom with no religious foundation.
Amor proprio - self-esteem
Anting-anting - Moro charm or amulet. Good luck charm
Balicuate - (slang) evacuate
Banca - small boat, very common
Barong - a large two-handed sword of the Moros
Barrio - a small native village
Carabao - water buffalo
Cargadore - porter
Dalama - boat
Datu - Moro title; community leader
Hapons - (slang) Japanese
Hiya - a sense of shame
Ilustrados - community leaders, land owners
Juramentado - frenzied, well-prepared ceremonial killing; Moro religious rite.
Kaingineros - nomadic gardeners
Kalibapi - political party of the Philippine puppet government
Kempeitai (Japanese) - military police
Kris ý Moro - wavy-edged knife
Mestizo, mestiza - mixed blood
Nipa - palm; leaves of which are used to make houses
Pakikisama - desire to avoid placing others in a stressful situation.
Paltik - homemade shotgun
Peso - Philippine currency
Sacada - person captured for forced labor
Sultan - temporal and spiritual leader of the Moros
Suyoks - sharpened bamboo stick placed in ground for a trap
Tankong - fern greens

Taos - fanners, laborers, share croppers
Tapa - beef jerky made from water buffalo meat
Tinghoy - counterfeit, useless (refers to guerrilla currency)
Tuba - coconut beer
Tulisan, tulisaffe - (slang) - thief
Utang na loob - denotes primary debt, reciprocal obligations
Voluntarios - home guard
Zona – Zonification, Japanese terror tactic

Dramatis Personae

South West Pacific Area (SWPA)

General Douglas MacArthur	Supreme Commander, SWPA
Colonel Charles Willoughby	Chief of Intelligence, SWPA
Colonel Courtney Whitney	Chief of AIB
Commander Charles Parsons	Cdr, SPYRON
Major Charles Smith	SPYRON

USFIP (10th Military District)

Colonel Wendell Fertig	Cdr, USFIP
Lt. Colonel William Bowler	Cdr, "A" Corps, USFIP
Lt. Colonel Hipolito Garman	Cdr, 105th Division
Lt. Colonel Frank McGhee	Cdr, 106th Division
Lt. Colonel Claro G. Laureta	Cdr, 107th Division
Lt. Colonel Charles Hedges	Cdr, 108th Division
Lt. Colonel James Grinstead	Cdr, 109th Division
Lt. Colonel Ernest McClish	Cdr, 110th Division
Major Herbert Page	Cdr, 116th Inf, 106th Division
Captain Donald LeCouvre	Cdr, 121st Inf, 105th Division

Filipinos

Manuel L. Quezon	President of the Commonwealth of the Philippines
José P. Laurel	President of the Second Philippine Republic, a Japanese puppet government established in 1943.

Manuel Roxas

First President of the Independent Third Philippine Republic, and clandestinely an Allied spy buried deep in the puppet government.

Japanese

General Masaharu Homma   Cdr, IJA 14th Army

General Tomoyuki Yamashita   Military Governor and Cdr of all Imperial Japanese forces in the Philippines (replaced Homma).

General Sōsaku Suzuki Cdr, IJA 35th Army (Mindanao) 1944

Lt. General Jiro Harada

Cdr, 100th Division, created in Mindanao in June 1944.

Lt. Gen Gyosaku Morozumi

Cdr, 30th Division which defended Mindanao during the US invasion.

General Iichiro Morimoto

Cdr, Japanese Occupation Forces Mindanao (1942-43), Cdr, Philippine Islands POW camps

Colonel Yashinari Tanaka

Cdr, Western Mindanao (1943)

Maps

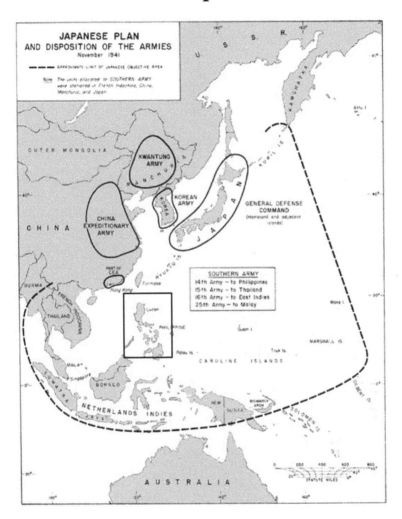

Japanese Plan for Asia (Courtesy of US National Archives)

Map of the Philippines

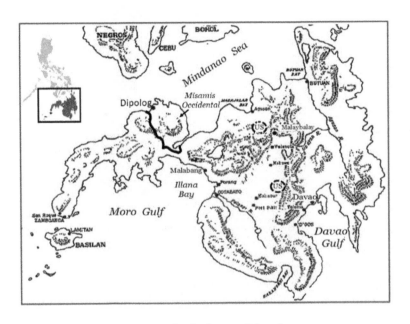

Map of Mindanao Island

Colonel Wendell Fertig
16 December 1900 – 24 March 1975

Introduction

"Above all, the most characteristic feature of insurgency in general will be constantly repeated in miniature: the element of resistance will exist everywhere and nowhere." — Clausewitz

After reading thousands of pages of literature on the Philippine resistance and the exploits of the Filipino guerrillas, a simple observation finally hits the reader: the faces of the prominent characters of the resistance remain a mystery because there are virtually no photographs of the guerrillas. Almost none of the personal accounts have pictures of any kind, and few of the secondary sources do. This is symptomatic of the general phenomenon, which is, the Philippine resistance did not have its chroniclers moving with the guerrillas to detail its adventures and accomplishments. It is understandable why there were no cameras or photographers. But this one small observation underscores the reason why so little is known of the resistance of the Filipinos to the Japanese.

Several authors have alluded to the fact that the definitive book on the Philippine resistance movement has yet to be published. Whereas historians have been provided with multitudinous accounts and analyses of the European resistance movements and the general public satisfied with novels and movies showing the daring of these resistance fighters, few in the West are even aware of the Filipino's struggle against the Japanese.

That is not to say that accounts of the Philippine resistance do not exist - they do, and in large numbers in the Philippines. The void lies in English language accounts available in the West. There are some English language accounts available, but judicious use must be made of them in order to create an accurate picture of the events in the Philippines during the resistance.

Accounts of the Philippine resistance generally focus either on the Japanese treatment of prisoners in the internment camps or upon the collaboration issue. Personal accounts written by or about members of the various guerrilla organizations for the most part do not deal adequately with the problems of guerrilla organization, logistics, relationship to the civil government, tactics or the politics of the guerrilla resistance. The various accounts - including the guerrilla unit histories tend to be self-serving, short on facts, and exercises in ax-grinding. But this problem is surmountable, given enough sources from which to glean information and make comparisons.

Many diaries were kept by guerrillas, and some very good personal accounts have been written from them. Official documents can sometimes serve as the catalyst to extract the truth from this information. Nevertheless, there is no one book which is a scholarly, in-depth account of the Philippine resistance movement as a whole, and there is no exhaustive record published on the guerrilla organization on Mindanao.

Research for this book relied heavily upon personal accounts and upon United States Government documents. Because little was known of their activities during the war, government documents provide information on the guerrillas sparingly. The personal accounts are subject to the criticisms already given. One untapped source for information on the Philippine guerrillas which was used for research in this book is the Philippine Archives in the National Personnel Records Center in St. Louis, Missouri.

The Filipino-American resistance to the Japanese on the Island of Mindanao remains one of the most successful unconventional warfare (UW) campaigns in US History. In order to demonstrate this, we must first define our terms and lightly trace out the contours of what is known as insurgency theory.

As current US doctrine holds, there are specific physical and environmental conditions that allow for a successful resistance or insurgency. The three main conditions are: (1) a weakened or unconsolidated government or occupying power; (2) a segmented population, and (3) favorable terrain from which an element can organize and wage subversion and armed resistance.[1] As this study will show, the Mindanao resistance capitalized on these conditions.

The bedrock of these conditions stem from several weighty, time-honored minds. First, the analysis of Carl von Clausewitz, in his treatment of what he called "the people's war," offers the following five conditions to creating an effective uprising:

1. The war must be fought in the interior of the country.
2. It must not be decided by a single stroke.
3. The theatre of operations must be fairly large.
4. The national character must be suited to that type of war.
5. The country must be rough and inaccessible, because of mountains, or forests, marshes, or the local methods of cultivation.[2]

In the last analysis, for Clausewitz, the most characteristic feature of an insurgency is ubiquity; to be

---

[1] TC 18-01, *Unconventional Warfare* (Washington, D.C.: Government Printing Office, 2011), 1-3.
[2] Carl von Clausewitz, *On War, Everyman's Library* (Princeton, NJ: Princeton University Press, 1993), VI.26, 578.

nebulous and elusive; to "exist everywhere and nowhere."[3] As will be seen, Fertig's Mindanao resistance followed, albeit most likely unconsciously, each jot and tittle of Clausewitz's prescription.

A century later, in his book *Seven Pillars of Wisdom*, T.E. Lawrence evoked his six principles of guerrilla warfare:

1. A successful guerrilla movement must have an unassailable base.
2. The guerrilla must have a technologically sophisticated enemy.
3. The enemy army of occupation must be sufficiently weak in numbers so as to be unable to occupy the disputed territory in depth with a system of interlocking fortified posts.
4. The guerrilla must have at least the passive support of the populace, if not its full involvement.
5. The irregular force must have the fundamental qualities of speed, endurance, presence and logistical independence.
6. The guerrilla must be sufficiently advanced in weaponry to strike at the enemy's logistics and signals vulnerabilities.[4]

Again, as it will be seen, the success of the US-led resistance on Mindanao was mainly because it met Lawrence's conditions.

Following Lawrence, the next insurgent theorist who plays a primary role is Mao Tse-Tung. For Mao, the underlying cause or ideology in an insurgency is of the utmost concern. In other words, a cause "is" an insurgency. Among Mao's contributions is his three-phase approach to how insurgencies mature. As outlined in his book *On Guerrilla Warfare*, Mao's three-phase

---

[3] Ibid., VI.26, 581.
[4] Thomas E. Lawrence, *Seven Pillars of Wisdom* (New York: Anchor, 1991), 196.

construct is a useful template for planning and developing an insurgency. These three phases are: latent or incipient, guerrilla warfare, and war of movement.[5]

As to cause and developmental phases, the US-led resistance on Mindanao once again is quite prototypic. And it may be conjectured, with a high degree of certainty, that Fertig was neither familiar with the history or philosophy of guerrilla warfare. Moreover, he probably had not read Lenin and certainly not Mao Tse-tung. However, using common sense, he was aware that the first essential element to a resistance movement was a cause.[6] And as will be seen, a cause is the most powerful when it can combine multiple facets.[7]

As this study will show, the Mindanao resistance is prototypical to the ideal resistance movement. In this vein, the purpose of this study is threefold: First, to demonstrate how the US-led resistance was successful against the Japanese occupation force, which was primarily for seven reasons:[8]

1. The ability of Fertig and his subordinate leaders to unify disparate groups under one unified command (leadership).
2. The ability of Fertig and his subordinate leaders to organize the resistance and give it vision and direction (ideology).
3. The UW tactics employed (objectives).
4. The terrain on the Island of Mindanao was favorable to UW (environment and geography).
5. The ability of MacArthur's command (HQ, SWPA) to legitimize and resource the resistance (external support).

---

[5] TC 18-01, *Unconventional Warfare*, 2-6.
[6] Robert B. Osprey, *War in the Shadows: The Guerrilla in History*, Vol 1 (New York: Double Day, 1975), 522.
[7] David Galula, *Counterinsurgency Warfare: Theory and Practice* (Westport, CT: Praeger, 2006), 15.
[8] These seven reasons correspond to the seven dynamics of a successful resistance movement.

6. The popular support and resources of the Filipino people and President Quezon; auxiliary and underground (internal support).

7. The ability of Fertig and subordinate leaders to adapt their tactics in support of the US invasion (phasing and timing). See diagram below.

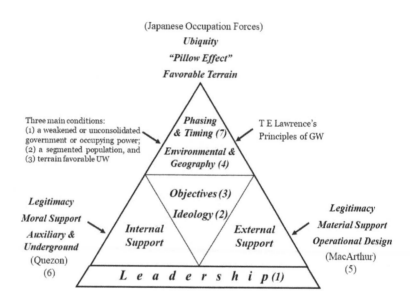

(Japanese Occupation Forces)
Ubiquity
"Pillow Effect"
Favorable Terrain

Three main conditions:
(1) a weakened or unconsolidated government or occupying power;
(2) a segmented population, and
(3) terrain favorable UW

Phasing & Timing (7)

T E Lawrence's Principles of GW

Environmental & Geography (4)

Objectives (3)

Ideology (2)

Legitimacy
Moral Support
Auxiliary & Underground
(Quezon)
(6)

Internal Support

External Support

Legitimacy
Material Support
Operational Design
(MacArthur)
(5)

L e a d e r s h i p (1)

Seven Dynamics of a Successful Resistance Movement

Second, to analyze how the Japanese failed in their counterinsurgency efforts against the Mindanao resistance. And third, to demonstrate the overall effectiveness of General MacArthur's UW campaign.

The intent is to validate Fertig's approach as a model resistance movement, within the context of MacArthur's prodigious UW campaign on the one hand, and Japan's bungled counterinsurgency campaign on the other. The goal will be to trace out these three strands of thought and balance them "like an object suspended between three magnets."

Analyzing this, the chapters of this study alternate between the above stated three strands of thought as follows: Chapter one canvasses the influence of the Japanese on the Philippines before the war, as well as the background of the resistance. Chapter two will discuss the Japanese invasion, along with the initial contact and infiltration of unsurrendered warfighters.

From a counterinsurgency (COIN) perspective, chapter three focuses on Japan's intent to force compliance on the Filipinos. Chapters four to twelve highlight the organization and build-up of Fertig's resistance movement on Mindanao, followed by its operational employment in preparation and support during MacArthur's invasion in chapters thirteen to fifteen. Finally, chapter sixteen draws conclusions from the discussion presented in the book.

Not long after the Japanese invasion, Filipinos who opposed the Japanese occupation began an underground organized resistance. Ultimately, this guerrilla activity encompassed the whole archipelago until more than 260,000 Filipinos were active in guerrilla operations to resist the Japanese. The major center of this resistance was on the Island of Luzon. Gradually other guerrilla units were formed on other outlying islands. In time, these disparate guerrilla bands would be consolidated under McArthur's United States Armed Forces Far East command. This book is primarily the story of the guerrilla resistance on the Island of Mindanao under the command of Army Lieutenant Colonel Wendell Fertig.

Chapter One
The Commonwealth of the Philippines

"The prospects and progress of a guerrilla movement depend on the attitude of the people." – B.H. Liddell Hart

The Filipino people have a long history of uniting against foreign incursions. In 1521, Ferdinand Magellan led a Spanish expedition to the Islands but was killed in a battle with native warriors several weeks later and his fleet departed. In 1542, the Spanish explorer Ruy López de Villalobos named the islands of Leyte and Samar *Felipinas* in honor of the Prince Philip of Asturias who would later become the King Philip II of Spain. In time, the whole archipelago would be known as *Las Islas Filipinas*. The first permanent Spanish settlement was made in 1565 by General Miguel Lopez de Legazpi. The Spaniards divided the Archipelago among themselves and employed the Filipinos as tenant farmers and servants. Spanish priests also converted most of the Filipinos to Roman Catholicism. Centuries of Spanish rule brought with it the Spanish language and culture.

While the Philippines prospered economically under Spain, a growing number of Filipinos sought to rid themselves of the Spanish colonial yoke. In 1892, a Manila clerk, Andres Bonifacio, formed *Katipunan*, a secret revolutionary movement.[1] Despite Katipunan's popularity, it was squashed by the Spanish colonial government. On August 19, 1896, hundreds of Filipino suspects, both innocent and guilty, were arrested and

---

[1] The word Katipunan literally means "association," and was short hand for Kataas-taasan, Kagalang-galangan, Katipunan ng mga Anak ng Bayan, which in English is "Supreme and Honorable Society of the Children of the Nation."

imprisoned for treason. Bonifacio and many other Filipino leaders were executed. Emilio Aguinaldo then became leader of Katipunan. To quell the nascent rebellion, Spanish colonial officials promised political reforms if Aguinaldo ended the revolt and left the Philippines.

In 1898, the United States and Spain went to war. In the aftermath, the US paid Spain $20 million for the Philippines and the Islands became an American territory. However, many Filipinos, including Emilio Aguinaldo claimed that the US had promised the Philippines independence. To precipitate matters, Aguinaldo declared the First Philippine Republic on January 23, 1899, beginning what would become known as the Philippine Insurrection (1899-1902). Aguinaldo's troops began fighting the Americans on February 4.

While Filipino nationalists viewed the conflict as a continuation of the struggle for independence that began in 1896, the US government regarded it as an insurrection. While details of the Philippine Insurrection are beyond the scope of this study, suffice to say that the cost of the conflict in terms of human lives was 4,200 American and over 20,000 Filipino combatants, and as many as 200,000 Filipino civilians who died from the ravages of war, famine, and disease. After Aguinaldo was captured by US forces in 1901, Filipino war sentiment withered and the fighting soon ended.

The US then set up a colonial government in Manila and Howard Taft, who later became the President of the United States, served as the first governor of the colony. During the period of American rule, the US introduced widespread literacy, improved public health, promoted an expanding prosperity throughout Filipino society, established free speech, increased civil liberties, and founded a representative government that drew upon a people with an increased sense of opportunity.[2] And as American business in the Archipelago increased, the Philippine economy became dependent on the United States. Moreover, the US began to allow Filipinos to serve in government. Then in 1935, the Philippines became a commonwealth with its own elected government and constitution, and Manuel Quezon became its first president. As the Philippines moved toward independence, the US retained authority mainly in foreign affairs as well as the defense of the Archipelago.

By the late 1930s, Japanese activities in the Far East had become sufficiently threatening to US interests. With the winds of war brewing in the Pacific, President Roosevelt recalled retired Army General Douglas MacArthur back into service, giving him command of what become the United States Army Forces in the Far East (USAFFE). The USAFFE included a strength of some 22,000 men, including 11,937 Philippine Scouts. However sizable, the fighting quality of this force was in question.

General MacArthur saw his immediate tasks as follows: First, establish his headquarters and organize his command; second, induct and train the Philippine Army; and third, secure the necessary supplies and reinforcements to bring the USAFFE to a war footing. For this incredible task, MacArthur was given $10 million to fortify the Philippines against the looming attack of Japan. Further, on July 26, 1941 a Philippine

---

[2] Rafael Steinberg, *Return to the Philippines* (New York: Time Life, 1980), 25.

Presidential Order transferred all organized military forces of the Commonwealth into USAFFE.

Meanwhile the War Department helped MacArthur by flying him thirty-five new B-17 Flying Fortresses, which was one-third of the existing US bomber strength. It was believed that these bombers could prevent an attack on the Philippines.[3] However impressive these bombers were, the large Filipino force which made up the bulk of MacArthur's defense was handicapped by poor training, virtually non-existent supplies, obsolete weapons and military equipment, no artillery, and inadequate leadership. As it turned out, the $10 million barely built a few training camps and induction centers.

To further compound the problem, President Quezon deemphasized the defense program during the two years preceding the Japanese invasion. Quezon naively believed that the Philippines was neither economically or military important to Japan. Thus, he hoped to steer his country on a course of neutrality in the event of war between Japan and the United States. Despite it all, optimism was high. MacArthur exuded great confidence in himself, his staff and the untried Filipino soldiers.

The American-administered Commonwealth of the Philippines also employed a host of Americans from various backgrounds. One of these was Wendell Fertig. Fertig came to the Philippines five years before the invasion during the mining boom of the 1930s with many other American engineers who had heard of great untapped gold and coal resources. Tall with an athletic build, Fertig was born in the small town of La Junta,

---

[3] John D. Lukacs, *Escape from Davao: The Forgotten Story of the Most Daring Prison Break of the Pacific War* (New York: Simon & Schuster, 2010), 11.

Colorado, which in 1900 was every bit a frontier western town.

Attending the University of Colorado, majoring in chemistry, Fertig transferred to the Colorado School of Mines in Golden, Colorado to be a mining engineer. While there, he enrolled in the US Army Reserve Officer Corps. After graduating in 1924, he was commissioned a reserve lieutenant and married his wife Mary. Then, in 1936, he moved his family to the Philippines where he pursued a career as a civil engineer. His first job was an on-site supervisor of a start-up mine in the Province of Batangas, south of Manila. Later, Fertig moved to Manila to take on a consulting position. From which he made several business trips to Japan. At the time, Japan was a major importer of the minerals from the mines Fertig supervised.

As witnessed by Fertig, in the decade preceding Japan's invasion, the Japanese had established strong economic ties with the Filipinos. The focus of this activity centered on Davao Province in Mindanao. In fact, in the period 1930 to 1939, some 19,000 Japanese immigrants came to the Philippine Islands; averaging over 2,000 a year. Eighty percent of the immigrants settled in Davao.

The Japanese immigration to the Philippines was nearly ubiquitous as they settled in nearly every province. Moreover, every town had at least one or two Japanese nationals in it.[4] It may be conjectured that the Japanese economic penetration of Davao began in 1907. This occurred when the Ohta Development Company imported 150 Japanese laborers to work in the abaca

---

[4] In fact, with some 29,000 Japanese in the islands, 64 percent (18,733) were living on Mindanao, and approximately 95 percent of those on Mindanao lived in Davao Province (17,888 residents).

fields. This early foothold grew into Japanese interests in shipping, fishing, lumber, and in the iron, manganese and copper mines. The immigrants established their own schools, newspapers, stores and banks. Moreover, the Japanese owned 70 percent of the hemp produced in Davao, and controlled the remainder through one means or another. All told, the Japanese investment in the Philippines was over 64 million pesos (32 million dollars), 50 million pesos of which was invested in Davao.[5]

Japanese ownership of the Philippines caught the attention of the Philippine central government, albeit slowly. To stem the Japanese investments, the Commonwealth government passed the Public Land Act of 1936 which required that at least 60 percent of the capital of any corporation dealing with the public be owned by Filipinos. The Japanese circumvented this law, and by 1939 Japanese investors owned between 142,000 and 148,000 acres of land in Davao of which only 70,000 acres were legally acquired.[6]

Further, Japanese investors circumvented the Public Land Law by establishing dummy corporations. Women from the interior tribes were purchased as wives for Japanese investors. Land was then purchased in their names, and then they were returned to the tribes in the mountains. It became a commonly held view, that once the Philippines would eventually gain independence from America, the Japanese would by that time gain political and economic control. Slowly, Japan built up a fifth column in the Philippines, employing the now

---

[5] Catherine Lucy Porter, *Crisis in the Philippines* (New York: Porter Press, 2012), 99, 102.
[6] Teodoro A. Agoncillo, *The Fateful Years: Japan's Adventure in the Philippines, 1941-1945* (Manila: University of the Philippines Press, 1965), 48-49.

familiar technique of commercial penetration.[7] All of this would fit well into the fold of the future Japanese philosophy of the Greater East Asia Co-Prosperity Sphere. However, because the Japanese were good citizens who brought in good business, most Filipinos weren't really alarmed. The same could not be said for the Philippine government or the USAFFE.

With the prospects for war with Japan growing daily, and the Army needing engineers, Fertig was brought on active duty on June 1, 1941 with the rank of major. His first assignment involved overseeing the preparation and improvement of airfields throughout the archipelago. Then in July, his family left for the United States on the *President Taft* along with the last of the Army families.

Describing the state of the USAFFE in those days, Fertig observed the soldiers were very proficient at close order drill, but little real training was accomplished beyond that. There were no adequate ranges, and no money to build them. Training ammunition was limited to ten to twenty rounds per soldier for marksmanship and familiarization training.

Moreover, the Philippine Army simply had no supplies to issue to the soldiers. What it did have was either old or did not fit. In fact, most soldiers were barefooted and lacked clothing. That the Filipinos did not have the M-1 was no surprise, of course. At the outbreak of World War

---

[7] A fifth column is any group of people who undermine a larger group from within, usually in favour of an enemy group or nation. They may work in an army, political party, or industry. The activities of a fifth column may be overt or clandestine and consist of spying, sabotage and propaganda. The term "fifth column" comes to us from the Spanish Civil War (1936-1939) and describes the work of Francisco Franco's followers in Madrid. Emilio Mola, a leader under Franco said, "I have four columns moving against Madrid, and a fifth will rise up inside the city itself."

II, the US Congress had still not authorized their production in any quantity for US forces which were soon to be in combat with the Germans and Japanese.

The basic infantry weapon was the old Enfield rifle which had defective extractors and a stock which was too long for most Filipino soldiers. As for machineguns, the .50 calibers were obsolete and lacked necessary water-cooling devices. The .30 caliber machineguns were unserviceable, and there were no replacement parts for them or any of the other weapons. Moreover, there were no anti-tank guns, hand grenades, gas masks or steel helmets, and the signal equipment did not work.

The peaceful archipelago would soon be the stage on which one of the world's greatest dramas would play out. And the USAFFE would have a front row seat.

Chapter Two
Run to the Hills!

"Thunder clouds of this type should build up all around the invader the further he advances. The people who have not yet been conquered by the enemy will be the most eager to arm against him; they will set an example that will be gradually followed by their neighbors."– Clausewitz

Seeking hegemony and the immense natural resources in Asia, particularly the Pacific rim, Japan's strategic objectives included the subjugation of the Philippines along with the capture of Malaya and the Dutch East Indies. The Philippines are ranked among the three richest island groups in the world. Moreover, with the Philippines in hand, the Japanese could follow up with the acquisition of Borneo's rich oil fields. To this end, the conquest of the Philippines became an immediate military necessity. On December 8, 1941, the day after their attack on Pearl Harbor, the Japanese launched an invasion of the Philippines by sea from Formosa (Taiwan). Over the next few days, the Japanese made landings at Bataan, Vigan, Aparri, Legaspi, Davao, and Jolo (See figure 2-1).

To the south at Mindanao, a Japanese force landed at Davao on 20 December. It consisted of two groups, the Sakaguchi Detachment and Miura Detachment. Arriving just after midnight, these detachments quickly overran the American-Filipino defenders, and by 3:00 pm that same day, Davao and its airfield were occupied. The first assault on Davao was aided by numbers of Japanese Fifth Columnists.

After taking Davao and its airfield, Sakaguchi's Detachment (56th Division of the IJA 16th Army)

departed for the Dutch East Indies while the 1,200-man Miura Detachment remained at Davao. The Miura Detachment, led by Lt. Col. Toshio Miura, consisting of 1st Battalion, 33d Infantry, were attached to Homma's 14th Army and became the beginnings of the occupation force for Mindanao (See figure 2-1).

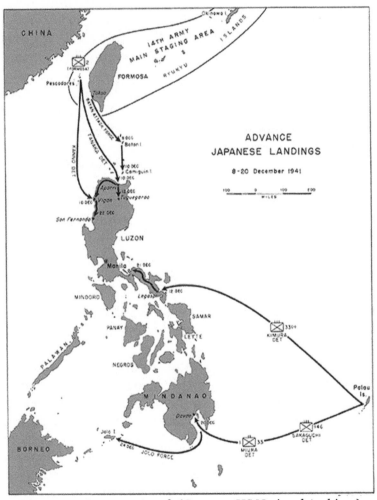

Figure 2-1. Japanese Attack (Courtesy US National Archives)

This sudden, dramatic and wholly unexpected Japanese attack found the 30,000-man USAFFE totally unprepared. At Clark Field, some fifty miles north of Manila, the American bombers and fighters were caught on the ground and most were destroyed. "MacArthur's air force had suffered a death blow."[1] Consisting of a mixed force of regular, national guard, and newly created Commonwealth units, the defenders were no match for the 43,000-man 14th Army of the Imperial Japanese Army (IJA) under Lieutenant General Masaharu Homma.

On Luzon, with Japanese forces bearing down on Manila, MacArthur ordered the USAFFE to fight a delay action. After his troops retrograded across Bataan to Corregidor, under Washington's orders, MacArthur left for Australia on February 22, 1942 along with Filipino President Manuel L. Quezon. Before departing, MacArthur divided the USAFFE into four separate commands, all co-equal and subordinate to him, among which was General Jonathan M. Wainwright's Luzon Force and Sharp's Mindanao Force. However, he failed to inform the War Department of this reorganization, and later both General Marshall and the conquering General Masaharu Homma believed all American forces in the Philippine Islands were effectively under Wainwright's command.

On April 9, 1942, following three months of savage fighting, Bataan fell to the Japanese. The valiant defenders, with no air or naval support, ran out of supplies and ammunition but not guts. The Japanese conquerors marched into Manila. To avoid its destruction, the Philippine President Manuel Quezon declared it an "open city." After the fall of Bataan, over

---

[1] Lukacs, *Escape from Davao*, 20.

60,000 American and Philippine prisoners were forced by the Japanese to undertake the infamous "Death March" to prison camps some 105 kilometers to the north. Before reaching their destination, it is estimated that some 10,000 men, weakened by disease and malnutrition, endured unthinkably harsh treatment by their captors, until succumbing to death.

Meanwhile at Davao, Colonel Miura had attempted to extend his control into the interior of Mindanao but was without success. To assist him, Homma dispatched the Kuwaguchi Detachment which made landing on April 29, 1942 near Cotabato (See figure 2-2).

With a force of 4,800 men, and some assistance from the Miura Detachment, General Kawaguchi then pushed back Brigadier General Guy Fort's 81st Division (part of Sharp's command) to gain control of the southern and western parts of Mindanao. Only in the north, in the Cagayan Sector, were the Filipinos still strong enough to offer organized resistance. Then Homma followed up with the Kawamura Detachment four days later at Cagayan Bay on May 3rd.

Figure 2-2. Island of Mindanao.

The period of May 6 through May 10 that followed was one of great confusion. This was brought on in part by a misunderstanding over command relationships and greatly complicated by the difficulties of communicating by radio and fighting the pressing Japanese attacks. With what was left of the USAFFE under Wainwright in the west, two major factors led to Sharp's decision to capitulate on Mindanao: the Japanese were fully capable of massacring the 10,000 survivors of the Corregidor garrison, and it was now clear that US reinforcements for the Mindanao Force would not be forthcoming as envisioned.

As William Breuer observes: Unbeknownst to any of the commanders on the ground, President Roosevelt and his Joint Chiefs of Staff "had written off the

Philippines, sacrificing the American and Filipino troops in the islands, along with 17 million natives, on the altar of global expediency in order to buy time for a woefully unprepared United States to rearm and build combat-ready forces."[2]

Still, few commanders in the south were so hard pressed as to be incapable of further resistance, and none had any desire to surrender. Elements of Sharp's Mindanao force were still undefeated, intact and capable of continuing an organized resistance. Plans had been made for withdrawal to the interior, and junior commanders awaited only the orders to execute the withdrawal. The Miura Detachment then moved northwest from Davao and made link up with Kuwaguchi on May 6. By May 9th the fighting was over.

Surrender

When Sharp had determined that further resistance would be fruitless, the subordinate commanders were ordered to surrender their men, weapons, and equipment to the Japanese. But whereas Sharp himself had little alternative to surrender, his subordinate commanders enjoyed somewhat more flexibility. While the actual number of Americans and Filipinos who took flight to the hills is unknown, the US Army concluded that the number who refused to surrender and disappeared into the hills of Mindanao probably exceeded those of all the other major islands combined.[3]

---

[2] William B. Breuer, *MacArthur's Undercover War: Spies, Saboteurs, Guerrillas, and Secret Missions* (New York: John Wiley & Sons, 1995), 12.
[3] Steve Mellnik, *Philippine Diary 1939-1945* (New York: Van Nostrand Reinhold, 1969), 281.

Father Haggerty, an American Roman Catholic priest, was with Sharp at his headquarters in the final days before the surrender. He put the Mindanao Force at an estimated 35,000 Filipinos and 1,000 Americans, the Americans being mostly headquarters and Air Corps personnel. Moreover, Haggerty observed, of that number, an estimated 7,000 entered the internment camps.

After Sharp's surrender on May 10th, "A deep, black pall of silence settled over the whole archipelago."[4] There were no radio transmissions. It was as though the Islands had been physically blotted off the face of the map.[5] Still, MacArthur was certain that there would be resistance, although what form and organization it would take no one could know. As Virgil Ney concluded in his two studies on guerilla warfare: "The source of the guerrilla idea was essentially and traditionally Filipino. To take to the hills and from there resist the invader has always been normal procedure in the Philippines, since the earliest Spanish days."[6] This phenomenon which had been tradition for the Philippines was something of an historical imperative on the island of Mindanao.

In the wake of the Japanese invasion, many decided to go home to wait and see what the new conqueror would do. Others did what the Mindanaons always did, they took to the hills to fight. But for the American and Filipino soldiers who were sworn to obey their orders, and who had made a career of doing so, the decision to surrender or not was much more difficult.

---

[4] Douglas MacArthur, *Reminiscences* (New York: Ishi Press, 2010), 202.

[5] Courtney Whitney, *MacArthur: His Rendezvous with History* (New York: Praeger, 1977), 128.

[6] Virgil Ney, *Notes on Guerrilla Warfare: Principles and Practices* (Washington, D.C.: Command Publications, 1961), 83.

Brigadier General Guy O. Fort, commander of the 81st Division, had vigorously contested Sharp's and Wainwright's orders to cease all resistance. Upon finally being persuaded that the surrender orders must be faithfully carried out, Fort admonished his troops: "As a soldier, I have no other alternative to follow but to obey orders. I expect you to do the same. No desertions will be tolerated."[7]

This order caused deep soul-searching for many, and most dutifully laid down their arms. The reasons some did not surrender are as varied as the circumstances and the number of men themselves. Captain Tom Jurika, brother-in-law of the famous Charles "Chick" Parsons, walked out the door as his commander held him at gunpoint to prevent his "desertion."[8] Many were ex-patriot Americans, like Jurika, whose adopted home was the Philippines; they held Reserve commissions, and had been called to active duty for the war. They saw no alternative but to continue the fight. Like Fertig, many believed that Wainwright and Sharp had surrendered under duress and therefore no longer had valid authority for the issuance of surrender orders. As for Fertig, he likened surrender to "castration."

For those who didn't surrender the first few months were the hardest. Unlike the Americans who had lived in the Philippines for many years, many soldiers had no knowledge of the customs, languages or terrain of the islands. A typical example was the 103rd Regiment which had defended the beach at Cagayan against the Kawamura Detachment. They had disengaged and

---

[7] Ira Wolfert, *American Guerrilla in the Philippines* (New York: Simon & Schuster, 1945), 142.
[8] Keats, *They Fought Alone: A True Story of a Modern American Hero* (New York: Turner, 2015), 10.

marched towards the Del Monte plantation expecting to find their headquarters there. With the surrender they disbanded. Soon individuals and small groups roved around without food. Many died of malnutrition and malaria.

Still others were set upon by roaming Moro bands and slaughtered.[9] Others, seeking refuge in the mountains, stumbled upon Moro villages or upon the pagan Magahats and were slaughtered.[10] Still others were made slaves of the Moros. Some, however, thought they had found nirvana as they were welcomed into tribal villages and were encouraged to impregnate all of the tribal chief's daughters so as to bring honor upon his house. Still others joined with the few American residents of Mindanao and established camps in the rain forests. One of these camps was presided over by Jacob Deisher, an American who had owned two sawmills and five mines near Illigan in Lanao Province.

When escape from the island was closed to him, he took $20,000, twenty trucks, fuel and provisions for two years and fled into the mountains west of Illigan. With some Spanish War veterans and his family, he set up camp, eating hogs, deer, and wild fruits and nuts. Fertig, who visited the camp, described it as a "wet hole in the jungle." He also reported that there were about thirty soldiers and sailors who wanted only to be left alone. They resented officers, would not take orders, and would do nothing but sit there, rotting in the jungle, living off the store of army rations, which Deischer, an old prospector and boar hunter, had gathered. As for

---

[9] Moros are Muslim Filipinos. The name "Moro" comes from the Spanish "Moor."
[10] Ibid., 167.

Deisher, he was just happy to have some company to ward off Moro raiding parties.

The first months after the surrender saw a total breakdown of authority throughout the islands. Luzon was shaken by renewed political rivalries and the emergence of the Communist Hukbalahaps (Huks). The islands were beset with rival political factions. As Colonel Allison Ind observes, there were at least six first-class wars going on not counting the official one on Leyte.[11] Following this trend, Mindanao saw the reemergence of the centuries-old feud between the Moros and the Christians, an especially bloody rivalry.

After the surrender, wild disorder prevailed as the Moros descended from the Lanao hills to plunder and pillage the lowland Christian settlements and to waylay the USAFFE soldiers making their way towards their homes. In the Malaybalay area, American and Filipino refugees were lured into apparent safety offered by the Lanao Moros, only to be killed. Throughout the war the trails in the area remained littered with hundreds of weather-whitened skeletons of the unwary.[12] As the level of violence increased, and the margin of security for virtually anyone outside of the few Japanese controlled areas plummeted to near zero, the island became an armed camp.

To make matters worse, the Philippine Constabulary, which before the invasion had maintained the peace, was now demobilized by the Japanese. Its 8,000 men were now placed in the same predicament as the Philippine Army. As the Moros raided and the lawless

---

[11] Hal Richardson, *One-Man War: The Jock McLaren Story, 1957* (Sydney: Angus & Robertson, 1957), 132.
[12] Ferdinand Miksche, *Secret Forces: The Technique of Underground Movements* (New York: Praeger, 1950), 14.

bandits fell upon the unprotected, USAFFE soldiers defended themselves, while the villagers and farmers were caught in the middle. If they kept guns, the Japanese would learn of their possession, and the penalty was beheading. If they turned their guns over to the Japanese as required to do, then they were at the mercy of the marauding bandits, who "requisitioned" anything, including women.

As for the Japanese on Mindanao, the combination of hill natives and guerrillas proved too much. And so, slowly yielding all but the coastal towns, they finally contented themselves to stop sending patrols into the interior of the island.[13] Ironically, this caused many to move to the coast to seek Japanese protection from their own countrymen. However, as is to be seen, the Japanese were not sympathetic.

As they say, "birds of a feather flock together." And so, the unsurrendered drew themselves together to form independent bands to protect themselves. This was for many the genesis of the Mindanao guerrilla movement. The reorienting of the thrust of the movement to resist the Japanese followed this initial phase.

In the vacuum of law and order, it was easy enough to start one of these early guerrilla bands. In addition to the need for security, the economic and social dislocation caused by the Japanese left many jobless. Among them were public servants, school teachers, truck drivers, boatmen, and former soldiers. If these men were not guerrillas, they were bums.

War brings out the very best and worst in everyone. And as armed resistance movements often attract 'bad hats,' unfortunately, the chaotic conditions did not facilitate the early rise of responsible leaders. In fact, the

---

[13] Osprey, *War in the Shadows*, 517.

situation gave some a license to indulge their vices and work off their grudges under the cloak of patriotism.[14] And so, the first leaders of these small guerrilla bands were the adventurers and desperadoes, the strong men and the passionate talkers.

Good, well-meaning men as well as brigands would follow these early leaders, and the reasons they did so were many. Many joined to avoid starvation, for an armed band could always acquire food. Still others were outlaws, driven by avarice. Many joined to settle old scores and feuds, and still others saw political gain in association with a particular leader. Thus, social conflict drove many into the ranks of the guerrillas. For example, on Luzon, joining the ranks of the Huks offered a chance to even the score with the wealthy landowners.

Finally, many joined for purely patriotic reasons out of allegiance to either President Quezon, MacArthur, the Philippine Commonwealth or America. The sentiments of these folks may be summed up by Colonel Marcos Agustin who so eloquently put it: "if the least we do is fertilize the soil where we fall, then we grow a richer grain for tomorrow's stronger nation."[15]

These early groups were separated by terrain and poor communications, and it would have been difficult to establish one cohesive group even had they so wanted. For although the majority of the guerrillas shared a common antipathy for the Japanese, they were often divided among themselves, separated into retractable rival factions engaged in a bitter struggle for power. Internecine strife and the struggle for power were part and parcel of this early guerrilla struggle. It was essentially the survival of the fittest; a distinct

---

[14] B.H. Liddell Hart, *Strategy* (New York: Meridian, 1991), 369.
[15] Agoncillo, *The Fateful Years*, 761.

evolutionary process to which all guerrilla units are subjected.

As for the conquerors, they were content to let this drama play itself out. As the divergent groups jockeyed for power, all the Japanese would have to do is wait in the hope that these miscreants would annihilate each other, saving them the trouble of doing the job themselves. Politics also played a hand. Some groups were bitterly opposed to the Quezon-Osmeña leadership in exile in America while others were ardent nationalists who were opposed to both Japanese and American hegemony. However, few of the guerrilla leaders were of the Philippine political elite. Many in fact were uneducated constabulary or Philippine Army officers who had no link to the post-war power structure; which itself was a continuation of the pre-war power structure.

As one observer noted, "As the guerrilla business grew down there (Mindanao) you had all sorts of people that were more interested in seeing who was going to be the boss of the island than in helping us."[16] This was to become a problem for Fertig and his American leaders, for it was perceived that after the war a successful guerrillero would be an obvious choice for public office. And the less he owed to the Americans the more obvious a choice he would be.[17]

Despite all the challenges, the indominable will of the Filipino people prevailed, and soon an organized form of resistance took shape. The development of the guerrilla movement on Mindanao mirrored that of the other islands in several respects. Initially, the guerrillas were

---

[16] Arthur Howard McCollum, *Reminiscences of Rear Admiral Arthur H. McCollum, US Navy (Retired)* (Annapolis, MD: United States Naval Institute, 1971), 598. A positive example is President Ravmond Magsaysay, who led guerrillas in the Zambales near the Bataan peninsula.
[17] Keats, *They Fought Alone*, 124.

formed under loose collections of individuals having common short-range goals. However, with the arrival of the second and third-rate Japanese garrison troops and the rise in banditry, the Filipinos joined together to restore civil order. By mid-August 1942, the various groups had consolidated under legitimate leaders and the spread of lawlessness was staunched.

As the guerrilla bands consolidated, three general categories of groups emerged. The first type gathered around a nucleus of unsurrendered USAFFE soldiers. A second category were derived from local pre-invasion leaders. A third group consisted of those with a pre-war social or political identification. The Moros on Mindanao and the Huks on Luzon both fit into this category.

As these guerrilla groups turned their attention to the Japanese, they got in each other's way, and lacking organization, exhausted themselves in uncoordinated actions. By February 1942, in order to make a whole of the fragments, MacArthur formulated a general concept for guerrilla operations which called for the disruption of Japanese lines of communication. In a preview of his "lie low" policy which he would issue a full year later, MacArthur wrote:

> I believe that extensive guerrilla efforts prior to the arrival of reinforcements from the United States will be abortive and destructive. I have complete plans to launch the guerrilla movement to support the main effort upon arrival of reinforcements.[18]

---

[18] Radiogram, *MacArthur to Marshall*, No. 178, 1 Feb 1942, AG 381 (21 Dec 1941).

Chapter Three
Co-Prosperity Sphere

"Violence takes much deeper root in irregular warfare than it does in regular warfare. In the latter it is counteracted by obedience to constituted authority, whereas the former makes a virtue of defying authority and violating rules."– Liddell Hart

During the course of the war, the Japanese understood rightly that their success in the Philippines was necessary if they were to expect to weld the Far East into an anti-Western bloc, the so-called Greater Far East Co-Prosperity Sphere. The intent of which was to "free" Asia from European colonial powers. However, as they implemented their policies, the Japanese demonstrated a twisted form of logic. Perhaps this is best expressed in the following proclamation which was published throughout the Philippines:[1]

> We have no intention of conquering any Asiatic people, nor do we have any territorial desire on any Oriental nation....But if you fail to understand the true and lofty purpose of Japan, and instead obstruct the successful prosecution of the military activities and tactics of the Imperial Japanese Forces, whoever you are, we shall come and crush you with our might and power, thus compel you to realize by means of force the true significance and meaning of our mission in the Far East.
>
> Commander in Chief of the
Imperial Forces
February 1942

---

[1] David Bernstein, *Philippine Story* (New York: Farrar, Straus and Company, 1947), 162. As John Tolland observes, the Co-Prosperity Sphere doctrine "had been created by idealists who wanted to free Asia from the white man."

The Japanese occupation policy in the Philippines was characterized by cruelty. The resentful Filipinos were quite unenthusiastic about their "liberation." What followed was economic exploitation, exposing the true nature of Japan's Co-Prosperity Sphere. Schools, businesses and small shops were closed. Utilities were turned off, transportation was shut down, and theaters, radios and newspapers banned. Within a year, some 1,000 Filipino businesses were taken over by Japanese businessmen. With the Co-Prosperity Sphere in full swing, the Filipino economic infrastructure supported the Japanese war effort. Mining equipment was dismantled and shipped away. Thousands of automobiles, trucks, and farm equipment were dismantled and shipped to Japan as scrap. Grain and rice harvests were seized to provide supplies for the Imperial Japanese Army (IJA). Those not seized were often burned.[2] Japan's Greater East Asia Co-Prosperity Sphere was anything but prosperity.[3]

This policy was especially damaging on Mindanao where the staple of the Filipino diet, rice, could be grown only in the coastal regions and river basins; areas controlled by the Japanese or easily accessible to them. The Japanese destroyed not only the mining and manufacturing infrastructure of the economy, they dismantled the support system for the peasant farmers as well by slaughtering carabaos for meat. This left farmers without the means to cultivate the fields.

It therefore goes without saying, the results of Japan's so-called Co-Prosperity Sphere policies were disastrous. In the wake of these polices, a high rate of unemployment and inflation drove prices sky high.

---

[2] Jack Hawkins, *Never Say Die* (New York: Dorrance, 1961), 166.
[3] Lukacs, *Escape from Davao*, 67.

Goods needed for daily living were no longer available. All of which led to a thriving black market. On top of it all, the production and import of medicines stopped, allowing disease to take hold on the population, in particular, malaria, pellagra, beriberi and typhus.

After the main fighting subsided, for their part, the Japanese did not have a particularly sophisticated pacification plan prepared for the islands. This lack of control kept the Philippines in a state of disorder. Perhaps one of their worst decisions, the Japanese disbanded the Philippine Constabulary Forces, which contributed to the chaotic atmosphere and left the population without a layer of protection from banditry and lawlessness.

The Japanese soldiers on occupation duty were also too few in numbers to effectively protect the population and their behavior contributed to the problem. Following the same trends, they took in China, Burma, Malaysia and other areas they conquered, the Japanese suppressed the Philippine population through terror. Rape, savagery, and murder occurred in every country the Japanese occupied. The looter's cage, hangings in conspicuous public places, and executions were common ways of terrorizing the Filipino populace into subservience.[4]

For their part, the Japanese administrators were unfazed by the degradation and carnage. In fact, they used prosperity as a weapon, rewarding collaborators with food while withholding it from those who didn't. Notwithstanding, the Japanese had not completely

---

[4] Charles Parsons, "Report on Conditions in the Philippine Islands as of June 1943," Intelligence Activities in the Philippines During the Japanese Occupation- 1943 (General Headquarters, Southwest Pacific Area, Military Intelligence Section, General Staff, June 1943), 1.

ignored the generally hostile attitude of the Filipinos. True to form, as they did with other guerrilla and resistance movements elsewhere, they gave the Filipino resistance little to no attention. In fact, a Japanese battle order for the Philippines directed soldiers to shoot guerrillas on sight.

The Japanese gave the guerrillas such short shrift because they perceived them to be mere bandits. Thus, counterguerrilla operations were called "punitive expeditions." To get their Asian cousins in the right mindset, the Japanese used radio broadcasts to assure them that all was well in the Co-Prosperity Sphere. For their part, Filipinos gathered in the barrio (village) plaza to listen to short-wave radio. At five o'clock P.M. every day, an Allied broadcast would be beamed their way. It would begin with General MacArthur intoning the words "I shall return," and end promptly at six o'clock P.M. with another promise to return.

Japanese propaganda belied the real struggle that was on-going between the guerrilla resistance and the Japanese forces. Along with propaganda, the Japanese used the full range of tools traditionally available to a government to fight guerrillas. They coordinated the activities of regular military units, national and local police, national intelligence and security elements, special police, agents, informers, and the public communications system. Bounties for guerrillas were paid to Filipinos in the form of food. For example, a bag of rice or meat would be exchanged for the head of a guerrilla.5 However, these tactics rarely worked, for few Filipinos would involve themselves with bribes, rewards or bounties, let alone sell out their countrymen.

---

5 Hernando J. Abaya, *Betrayal in the Philippines* (New York: A.A. Wyn, 1946), 57.

The savage war between the guerrillas and the Japanese was played out with no quarter given by either side. Guerrillas shot down Japanese whenever they could. Likewise, the Japanese would retaliate by placing the severed heads of guerrillas on stakes around the barrios as a warning to others not to join the resistance movement. Although less common on Mindanao than on other islands, guerrillas would ambush Japanese patrols on the outskirts of a town. The Japanese would then come and burn the town in reprisal and kill everyone in it. The rate varied, but the Japanese would exact a ratio of anywhere from 100:1 to 200:1 to an entire barrio for every Japanese soldier killed.[6]

And so, the civilians were happy to help the guerrillas if they would just confine their ambushes to the unpopulated areas. The guerrillas, in turn, saw no reason to invite the Japanese into their mountain sanctuaries.

As will be seen, the Japanese maintained control of the eastern third of Mindanao throughout most of the war. In this coastal area, the Japanese treated the population with uncharacteristic respect. As a result, Filipinos in this eastern third of the island were either on the side of the resistance or they sided with the Japanese.

Unbeknownst to the Japanese, this conciliatory approach to the people could have wrecked the resistance movement. As a principle there could be no neutral, middle ground. As will be seen, the guerrillas had to conduct sabotage and assassinate Japanese leaders in this area of Mindanao in order to provoke Japanese reprisals against these eastern towns. The

---

[6] Rafael Steinberg, *Return to the Philippines* (New York: Time Life Books, 1979), 32.

hope was that the Japanese, in doing so, would be exposed as the true enemy.[7]

In implementing the more brutal aspects of their occupation policy, the Japanese confronted, unwittingly or otherwise, the fundamental character of the Philippine culture. The Filipino holds sacred beyond all things his church, his family and his home. All these were systematically violated. It is said, the blood of the martyrs is seed. Characteristically, the church in the Philippines was strengthened during the occupation period.

On Mindanao, there were just forty priests to administer baptisms, conduct marriages, and preside over funerals. These clergymen became known as "guerrilla priests" for they travelled clandestinely with guerrilla units and were considered to be the most secure means of messenger communication.[8]

Father Edward Haggerty, who provides one of the best personal accounts of the Mindanao guerrillas, gives the reason why this was so: "I knew no Filipino would betray, above all, an American priest."[9] The priests would travel to remote areas to conduct services, hear confessions, and conduct mass marriages and baptisms. One step ahead of the Japanese, they served as a vital link which helped hold the island population together.

One might also conclude that because of their close relationship with both the guerrilla and civilian leaders they were able to encourage the guerrillas to be moderate in their treatment of Filipinos who had to live under the Japanese. This respect for priests reflects Filipino social values which, among other things,

---

[7] *They Fought Alone*, 197.
[8] The American, Canadian and Dutch priests joined the guerrillas.
[9] The one exception, Haggerty allowed, would be a "few Moros."

emphasizes personal honor, dignity, and pride. Of paramount importance within this framework, is the family which is the focus of social, economic, religious and, to a degree, political activity.

Within the framework of these values, it is easy to understand why various Japanese practices in the treatment of the population caused such a strong reaction from the Filipinos. For example, the Japanese treated the Filipino civilians much the same as they treated their prisoners of war. Civilians, regardless of station in the society, were required to bow to every soldier. Failure to bow could carry the severest penalties, such as being executed, trussed and hung in the barrio square or fried alive on galvanized iron in the hot sun.

The Japanese made one of their most grievous errors through their treatment of the Filipino women. Daughters were taken from their families to be prostitutes and concubines for the Japanese army.[10] When husbands and fathers would not cooperate with the Japanese, wives and daughters were raped and otherwise sexually abused before the man. For the Filipino, personal honor, and in particular a woman's honor, is sacred and can only be abused at the risk of one's life.[11]

As the seasoned Japanese combat units were reassigned elsewhere in the Pacific after the American surrender, poorly disciplined third-rate occupation troops, many of whom were Korean and Formosan conscripts, were brought to Mindanao to garrison the island.[12] As the abuse of the Filipino women and

---

[10] Carlos P. Romulo, *I Saw the Fall of the Philippines* (New York: Doubleday, Doran & Company, 1942), 253.
[11] Agoncillo, *The Fateful Years*, 648.
[12] Keats, *They Fought Alone*, 73.

personal wrongs inflicted became widespread, so did the number of men who had a debt to pay. Since one individual, family, or barrio could not pretend to deal with the Japanese Army alone, there was only one alternative – join the guerrillas as their means to repay the affront to their honor.

What remains to be seen is how the puppet Philippine government helped administer these Japanese policies. In the last analysis, the Japanese counterinsurgency (COIN) approach used in the Philippines is itself a case study on how not to do it. From collective forms of punishment to coercion and intimidation, nearly every poor COIN practice was employed. In fact, their approach served merely to bolster the ranks of various guerrilla bands.

Chapter Four
"General" Fertig

"If chance will have me king, why, chance may crown me Without my stir."– Shakespeare, *MacBeth*

Major General Charles Willoughby, MacArthur's Chief of Staff for Intelligence, pointed out the difficult problems confronting the guerrilla leaders and said "some emerged as really strong men, as leaders will always emerge in time of stress and disaster."[1] Fertig was one of these men, and as is to be seen, it was due primarily to his personal leadership qualities that the Mindanao resistance movement was unified and became the most successful of all of the guerrilla units in the Philippines.

Before the fall of Bataan and subsequent surrender, Fertig left Corregidor on the last aircraft to make it off of the island. Hugh Casey, MacArthur's Chief of Engineers, sent him to Mindanao to supervise the construction of airfields. Fertig was sent to the Dansalan-Illigan area to supervise the demolition of the main roads and bridges. That same day, the Japanese landed at Parang and Malabang and were pushing toward Lake Lanao. During the chaos, Fertig met up with Charles Hedges, an engineer friend whom he had not seen in years.

Then on April 29, 1942 the Kuwaguchi Detachment landed near Cotabato and the Kawamura Detachment followed four days later at Cagayan Bay on May 3rd. The Japanese landings sent Fort's 81st Division reeling in retreat. Then on May 10, Fertig learned of Sharp's surrender and heard that General Fort was still resisting.

---

[1] Charles A. Willoughby and John Chamberlain, *MacArthur 1941-1951*, 1954, 215.

His concern then was to link up with Fort's command until he learned of his surrender on May 27th.

From May through August Fertig remained in Lanao Province, living no more than a few kilometers away from the Japanese. Learning of his presence, the local Japanese commander, Captain Yamato sent Fertig a personal letter guaranteeing his safety if he would capitulate. Fertig sent no reply. Then on July 4th, Fertig and Hedges, from hiding, witnessed a "parade" of prisoners along the National Road from Dansalan (Marawi City) to Illigan.

The Japanese had placed General Fort in a truck bed at the head of the column of prisoners who were mostly barefooted and shackled together with wire. The "Independence Day Parade" had the sordid character of the Bataan Death March. Fertig dated his firm resolution to fight the Japanese from that day. From there, he withdrew into the mountains, waiting for the chaotic conditions on Mindanao to subside.

For the Japanese to control Mindanao would take quite a lot of effort. Mindanao is the second largest island in the Philippine Archipelago with 36,537 square miles, about the size of the state of Indiana. The island's coastline measures 1,400 miles and is defined by many gulfs, bays, inlets, capes, points and large peninsulas. The island has five major mountain systems with a varied and complex topography that includes numerous rivers and a number of lakes. Elevations in these mountain ranges run from 2,000 feet at the plateaus to a high at Mount Apo of 9,690 feet.

There are two major river systems. The Agusan River flows for 100 miles in the Davao - Agusan Trough from south to north on the east side of the island. In the southwest the Cotabato Lowlands surround the

Mindanao River which flows from east to west through large swampy stretches. The Bukidnon - Lanao Highlands with peaks rising to 9,500 feet contain the 134 square mile Lake Lanao. Fertig described this area as being "nearest to paradise. A high, cool, grassy plateau, cut by deep vertical walled canyons with blue-black mountains arising on all sides" which reminded him of New Mexico.[2]

As to terrain, it is rugged and inhospitable. In fact, the impenetrableness of the island's geography has been portrayed by the recent discovery of the Tasaday tribe. This primitive tribe had lived undetected in a rain forest among 200-foot trees in an unexplored area of Cotabato Province near Supu for an estimated 500 to 1,000 years. The irony is that this tribe, which has no word in its vocabulary for "war," was never accidentally discovered by the Filipino guerrillas who hid from the Japanese in this area. The existence of the tribe was reported in 1971.[3]

As to flora and fauna, the island had wild black bees that could kill carabaos (water buffalo) and pythons that would kill men. Both were common. Leeches lived on shrubs and in slimy streams. Swarms of mosquitoes infested the island, along with cobras, poisonous snakes, and crocodiles which dogged the footsteps of the unwary. Farmers were subjected to recurring locust plagues, and during the period of the Japanese occupation, vast hordes of rats left the forests to destroy entire rice crops. Not least of the hazards were inch-long rattan needles, bamboo spike traps set along trails for

---

[2] Wendell W. Fertig, "Guerrillero," Part I, Undated Manuscript, 24.
[3] Kenneth MacLeish, "Stone Age Cavemen of Mindanao," National Geographic, August 1972, 249.

wild pigs, deer, and humans, and headhunters who inhabited the interior.

Avoidance of these hazards was no guarantee of long continued health, however, for disease abounded. Yaws, filariarsis, scabies, tropical ulcers, worm infestations, dysentery, dengue, cholera, typhoid, scrub typhus and fungus infections were prevalent in varying degrees. The largest killer was malaria. Davao received rainfall of up to 11 inches a year, making the area a breeding ground for disease. The Mindanao terrain was, in short, ideal guerrilla territory.

Fertig was to write in his diary: "During the months in the forest, I have become acquainted with myself and developed a feeling that I do not walk alone; a feeling that a Power greater than any human power has my destiny in hand. Like a swimmer, carried forward by a powerful current, I can direct my course as long as my way lies in the direction of the irresistible flow of events. Never have I lost the feeling that my actions have followed a course plotted by some Power, greater than any human agency."[4]

Fertig felt himself destined for victory and part of a master plan. He was not recklessly messianic, however, for he counselled Hedges, who wanted to fight every Japanese he saw. In Fertig's logic, when the time was right, the Filipinos would come to them seeking their leadership. All he had to do is be patient and not force his leadership upon the Filipinos.

In this, he not only demonstrated an understanding of the ways of the Filipino culture, he also recognized an obvious truth, the Americans had been beaten by the Japanese, their leaders had surrendered or fled the islands, and the return of American fighting forces to the

---

[4] *They Fought Alone*, 104.

Philippines in the foreseeable future was not evidenced in any way. So Fertig awaited his opportunity. In late August he moved to Kolambugan and then on to Panguil Bay. As he saw it, his first task would be to recruit the unsurrendered.

On Mindanao, as elsewhere in the Philippines, the initial guerrilla organization centered around the particular personal qualities of various men. By September 1942, a former Constabulary junior lieutenant by the name of Luis Morgan had consolidated the small bandit groups in Misamis Occidental under his leadership (See figure 4-1).

An American Mestizo, Morgan had designated himself a captain and had joined forces with Lieutenant William Tait, a Mestizo son of a black Army veterinarian who had served with the old Negro Cavalry regiments in the Moro Pacification Campaign. Tait's exploits against the Japanese are legendary, but cannot be recounted here. His immediate superior, Morgan, was a natural leader with a great deal of charisma.

Described by Fertig as a virile, handsome, hard-drinking, absolutely fearless man, Morgan compelled the loyalty of men and the passion of women. It had been said among the Filipinos that "If a man is brave, and has a gun, he joins Morgan."[5] However fearless, Morgan could not provide the leadership necessary to consolidate a large number of guerrillas into a functional organization. To complicate matters, Morgan, a Christian, was locked in a continuing war with the Muslim Moros. Like so many others, in the wake of the American surrender, Morgan mercilessly attacked Baroy (15 kms west of Dansalan), and systematically massacred the Moro population. As it turned out,

---

[5] Ibid., 187.

Morgan became more of a "Moro killer, rather than a Jap killer."

Figure 4-1. Mindanao Island September 1942.

Morgan

On September 12, 1942, Morgan approached Fertig in the hopes of using him as leverage to assume authority over the entire island. Morgan offered his command of some five hundred Filipino guerrillas to Fertig on condition that he be made his chief of staff with authority to remain in the field. Morgan hoped that American leadership would consolidate the growing

Filipino resistance as well as pacify the Moros.[6] The courtship was not so simple. Perhaps the real beginnings of the guerrilla unification can be dated from then. Fertig summarized the negotiations with Morgan this way:

> Morgan found that he could not control the ambition of the various sector and area commanders. Any officer who had 12 rifles immediately appointed himself Major or Colonel. I realized that should this condition continue, internecine strife would result and the entire uprising result in a reign of terror for which the USAFFE would bear the stigma. It was decided that, in order to control these elements, I should assume the title of Brigadier General: this was done.[7]

In order to be taken seriously by potential recruits and rival guerrillas, Fertig promoted himself to brigadier general. His stars were fashioned from coins by a Moro silversmith. However, this self-promotion did not endear him to General MacArthur. In fact, MacArthur made it a point to send Fertig full colonel rank insignia along with an accompanying commission.

In Fertig's scheme, he would pose as a general sent by MacArthur to the Philippines to train a guerrilla army. This would give him immediate and absolute seniority over any contenders, for there had been no grade in the Philippine Army equal to a brigadier general of the US Army. For his part, Morgan saw in Fertig a solution to the seniority issue which would still leave him, Morgan, in charge. Unbeknownst to Morgan, however, Fertig would leverage Morgan's influence, and along with his knowledge of the Filipino people and the current

---

[6] Kent Holmes, *Wendell Fertig and His Guerrilla Forces in the Philippines: Fighting the Japanese Occupation, 1942-1945* (Jefferson, NC: McFarland, 2010), 27-30.
[7] *They Fought Alone*, 103.

situation, would eventually take over the command of that group and then others.

To that end, Fertig drafted his initial proclamation announcing his assumption of command on September 12, 1942:

UNITED STATES ARMY FORCES IN THE PHILIPPINES
OFFICE OF THE COMMANDING GENERAL IN THE
FIELD OF MINDANAO & SULU

18 September 1942

PROCLAMATION

On September 18, 1942, our forces under Maj. L.L. MORGAN completed the occupation of Misamis Occidental Province and Northern Zamboanga from the hands of the Japanese Military Government, and raised the American and Filipino flags therein.

In behalf of the United States of America, the Philippine Commonwealth Government is reestablished in those regions under the Military Authorities. All Civil Laws and regulations will be followed except in those cases where they conflict with Military Laws. In such cases Military Laws will prevail.

This procedure shall continue to be enforced until such a time when it shall be declared suspended, or terminated.

/s/ W. W. FERTIG
Brigadier General, USA.
Commanding Mindanao & Sulu Force.

As the proclamation letterhead indicates, Fertig considered his guerrilla organization to be a regular part of the US Army. Fertig's gambit, which proved to be the catalyst that would implement unity of command of the Mindanao resistance, had three effects: First, it gave Fertig the *bona fides* needed to begin talks with Moro clan leaders. Second, it provoked MacArthur to confirm rumors of a US Army general operating on Mindanao. And third, it prompted the unsurrendered to flock to his banner.

Soon after Fertig and Morgan came to an agreement, Fertig promptly ordered Morgan to travel to areas immediately bordering Misamis Occidental to encourage unification with neighboring guerrilla groups. Meanwhile, Fertig remained in the new headquarters in Misamis to confront the difficult administrative and logistical problems of organizing the resistance movement. Morgan was successful in Zamboanga and Sulu and returned to Misamis in December 1942.

Bamboo Telegraph

By runner and bamboo telegraph the word of Fertig's assumption of command was communicated throughout Mindanao. Guerrilla leaders rarely knew each other personally, thus many of the guerrilla chiefs greeted Fertig's messengers with skepticism or open scorn. Still, many soldiers, disillusioned after months in the jungle and sick from malaria, hunger and exposure, happily made the trek to Fertig.

And thus, Fertig started nearly from scratch. He was not familiar with the history or philosophy of guerrilla warfare, he probably had never read Lenin and certainly not Mao Tse-tung. But using common sense, he was aware that the first element essential to a resistance movement was a cause.[8] In his diary, Wendell Fertig wrote:

> I am called on to lead a resistance movement against an implacable enemy under conditions that make victory barely possible even under the best circumstances. But I feel that I am indeed a Man of Destiny, that my course is charted and that only success lies at the end of the trail. I do not envision failure; it is obvious that the odds are against us and we will not consistently win, but if we are to win only part of the time and gain a little each time, in the end we will be successful.[9]

By late September 1942, Fertig established goals for his organization and defined its mission.

To begin with, his tactical objective was limited to keeping his units dispersed and viable until US Forces returned. Corollary to this objective, was the need to obtain the unqualified support of the populace and establish a working civil government. To that end, he knew he had to effectively eliminate the bandit groups. With these gone, he could then persuade the civilians to give up their arms, which in turn would provide him with the weapons he needed to effectively engage the Japanese.

In this, however, Fertig faced a dilemma. In order to gain the support of the populace and subsequently recruit new guerrillas, Fertig had to demonstrate that he could, and would, fight the Japanese. On the other hand, he needed to preserve his forces and consolidate his administrative control in order to be effective. The way forward, as he correctly deduced, was by exercising caution, diplomacy, and selective guerrilla operations.

---

[8] Robert B. Osprey, *War in the Shadows: The Guerrilla in History*, Vol 1 (New York: Double Day, 1975), 522.
[9] *They Fought Alone*, 210.

In response to his proclamation, Fertig gathered a cadre of five officers and some 175 enlisted men.[10] He then worked at unifying the various scattered guerrilla bands, consisting mainly of four guerrilla organizations which had been founded by the Americans: Bowler, Grinstead, Childress and McClish. Bringing them into the fold, Fertig then appointed Robert Bowler as his second in command.

To fill his ranks, Fertig dealt with a curious phenomenon. Qualified experienced Filipino officers would ask him for an American to lead the Filipino officer's unit. One legacy of the American rule in the Philippines had been that the Filipino soldier still believed, even after the defeat of American arms, that Americans possessed superior wisdom. Fertig therefore found it necessary to commission junior enlisted men who had no leadership experience, no tactical training and no experience in order to attract the Filipino leaders he needed for his companies.

This infusion of young Americans into his guerrilla organization was not without its problems. The young Americans referred to the older former civilians as the Old-Timers Club. These Old-Timers were of course Fertig, Hedges and Wilson. Along with them were James Grinstead, an old lieutenant colonel and former member of the Philippine Constabulary turned planter; Major Herbert C. Page, well into his sixties, also a former planter; Cecil Walter, in his fifties, and Fred Varney, who had been a tough old mine operator, in his fifties as well.

Added to the group was Frank McGee, a retired officer who was decorated for action in World War I and who had lived on Mindanao for many years. In the middle

---

[10] Osprey, *War in the Shadows*, 518.

was Bowler and McClish, young officers, and therefore, presumably not lumped in with the Old-Timers.

Adding to the diversity, the Mindanao resistance included units which were sometimes all-Christian or all-Moro. There was also soldiers from the Philippine Army, Philippine Scouts, Philippine Constabulary, US Army, US Navy, US Army Air Corps, US Merchant Marine, former civilians, and perhaps even an Australian or two. As another ingredient, the influx of USAFFE soldiers into Mindanao before the surrender raised yet another problem for Fertig, the large variety of languages and dialects spoken. Soldiers from the Visayas and Luzon were on Mindanao as well as English-speaking Americans. In no time at all, Fertig had drawn an ethno-linguistically, religiously diverse, cosmopolitan crowd.

To conciliate and assuage various unharmonious issues, Fertig then sent Morgan off the island on a mission "to establish communications" with guerrilla groups in the Visayan Islands. During Morgan's absence, Fertig worked to reconcile his organization to the Church and the Moros. Morgan was in part the cause for this aloofness. The Church held Morgan to be bigamous and amoral, and the merchants and fishermen complained that he stole from them.

Fertig began with the Church. Knowing the Filipino people, Fertig understood that Doña Carmen, wife of former Senator Ozamis, was patroness of Misamis Occidental Province. Her word virtually carried the weight of law. If she supported his organization, the province would as well. Calculatingly, Fertig invited Doña Carmen to dinner, and wisely invited her priest Father Calanan, and her close adviser, Doctor Coureras as well. Fertig was persuasive at the dinner, and would later personally date the real birth of a unified guerrilla movement from this supper party.

Moreover, he calculated that the best way to convince the men to join his organization was through their wives, who were devout churchgoers. An ancillary benefit was that the wives would then work in the cottage arms industry to supply bullets and clothes for their men, whom they had in turn urged to join the guerrillas. Accordingly, Fertig ordered all guerrillas in Misamis to attend Mass. This demonstration of faith was also in stark contrast to the Japanese suppression of religious expression.

Catching word through the bamboo telegraph, Japanese intelligence began to refer to Fertig as "Major General Fertig. In fact, throughout the occupation period, Japanese intelligence referred to what would soon become the 10th Military District as the "10th Army Group." The Japanese commander of Mindanao and Sulu, General Iichiro Morimoto, was convinced that at this time Fertig had a force of some 7,000. Yet despite the fact that Morimoto was almost certain that Fertig's headquarters was at the old fort in Misamis City, because he had only three battalions of the 10th Independent Garrison on hand, his ability to launch an attack was limited.

Moreover, to ask Manila, by this time General Tanaka, for reinforcements would be unthinkable since Mindanao had long since been declared "conquered and pacified." To admit to problems on Mindanao now would prove him to be either an incompetent commander or of having been untruthful in his reports. To admit this would cause such a loss of face that he would, in all honor, be required to ceremoniously disembowel himself.[11]

Morimoto had fought guerrillas in China and had presumably believed that he did not have the forces available to deliver a crushing blow nor to adequately pursue, isolate, surround and destroy individual units as he had attempted to do tactically in China. His solution therefore was to create a special mobile combat team. He could piece such a force together by stripping down the garrisons to minimal strength.

If he could create a mobile combat team of significant force, greater than what he perceived Fertig's to be, then after obliterating Fertig, he could move this force to the next center of resistance and snuff it out.[12] Morimoto set out to piece together his counterguerrilla task force. But all this would take time, and time is always on the side of the insurgent.

---

[11] *They Fought Alone*, 127.
[12] Ibid., 126.

Chapter Five
The Krag and the Kris

"On the other hand, there must be some concentration at certain points: the fog must thicken and form a dark and menacing cloud out of which a bolt of lightning may strike at any time."
– Clausewitz

The cultural intelligence of Fertig and his subordinate commanders to understand the Filipino was to a great extent responsible for the welding of the resources of the Island into an effective military unit. As for the Moros, Fertig correctly understood to survive, the whole island would need to bury the hatchet and unite against the Japanese.

The Filipino Mohammedans (the Moros) were unique to the resistance movement on Mindanao. They had it within their ability to shift the balance of power to the Japanese or to cause an entirely different relationship among the antagonists by siding with neither. Sufficient numbers of the Moros actively supported the guerrillas or remained neutral. However, to have persuaded them to support the resistance was no small accomplishment.

In 1941, when war came to the Philippines, the Moros represented some twenty percent of the population of Mindanao and over ninety percent of the Zamboanga Peninsula. As Fertig saw it, clearly the Moros, as a group, represented a key segment of the island's population, and their influence was compounded by their naturally war-like bent.[1]

Within the Moro culture there are at least ten distinct groups differing in language and degree of Muslim religious orthodoxy. The four most prominent Moro

---

[1] *Crisis in the Philippines*, 103

groups are the Maguindanao, the Maranao, the Tausug and the Samal. Notably, while the Tausug are the most religiously orthodox, they consider the other Moros on Mindanao to be "uncouth."

These disparate groups, though known for their constant aggression against each other, will nevertheless close ranks when confronted with a threat from outside the Moro community. They feel a special spiritual orientation to the Islamic peoples throughout Southeast Asia, and they fear any group or invader who they perceive intends to deprive them of their religion and way of life. The Moros have preached separatism from well before the Spanish occupation of Mindanao, and that separatism carries through to today.

In approximately 100 BC, the Indonesian pagans, a Malayan-Mongoloid branch of the true Oceanic Malays, conquered the south Philippine islands. The Moro ancestry can be traced from these early conquerors. Perhaps apocryphally, Moro legends speak of a visit to Mindanao by Sinbad the Sailor. In fact, the Sulu Moros tell the legend of Alexander the Great holding court on Jolo Island in 320 BC. Many Moros still trace their descent from the great Macedonian prince.[2]

The Moslem influence came to Mindanao from Java and antedated the Spanish arrival in the Philippines in the 15th Century. For their part, the Spanish friars with their accompanying conquistadors tried unsuccessfully for four centuries to dislodge the Moro's deep-seated Islamic faith. Then, at the close of the 19th Century, the Americans supplanted the Spanish, but with a fundamentally different policy. The Americans were not particularly interested in converting the Moros from

---

[2] Vic Hurley, *Swish of the Kris: The Story of the Moros* (New York: E.P. Dutton, 1936) 35.

Islam but they were intent upon ending the violence perpetrated by them. It was during the Philippine Insurrection that the Americans employed tactics in their field operations and introduced firepower to the battle that the Moros had never experienced from the Spanish. That test was between the "Krag and the Kris," that is, the American Krag rifle (Springfield Model 1899 Krag) and the Moro's curved bladed kris.

The Moros found in the American an antagonist who made a game of war. It was during the Philippine Insurrection of 1899-1902, that the Americans would slay Moros on the weekdays and play baseball on Sunday. In fact, the vicious deadly guerrilla war was to last between the two enemies until March 22, 1915 when the Sultan of the Moros recognized without reservation the sovereignty of the United States over the Moro people. In return, the US promised the security for the Moros to practice absolute religious freedom.

The histories of the campaigns against the Moros are filled with stories of the nearly superhuman feats of Moros locked in mortal combat, and it is easy to understand why the Moro was so widely feared and their ferocity so legendary. No less legendary, were the experiences of John J. Pershing who served in Mindanao in three different tours to the Philippines. As a captain he had gone alone and unarmed into the "Forbidden Kingdom," the heart of the Moro land near Lake Lanao. No white man had ever entered the Forbidden Kingdom before, under any circumstances. Pershing had been ordered by Brigadier General Davis to "Do everything possible to get in touch with the Moros of central Mindanao and make friends of them."[3]

---

[3] Donald Smythe, *Guerrilla Warrior: The Early Life of John J. Pershing* (New York: Scribner, 1973), 65.

He believed that the only way he could truly demonstrate his good intentions was by putting himself entirely at their mercy. Pershing always carefully avoided the issue of slavery and polygamy, appealed to the Moros' self-interest, founded his policies in honesty and justice and resolutely avoided useless bloodshed. He led the expedition against the Moros at Bud Dajo, the Moro Sacred Mountain, where they made one of their last concerted stands against American arms. The slaughter was one-sided and great, and Pershing was to try later to refuse the Congressional Medal of Honor awarded to him for his role in this action.

In the eyes of the Moros, Pershing was a great warrior. And as such was made a Moro datu (tribal chief). Pershing's experience demonstrates that American policies had gained Moro respect and allegiance. So, there was a precedent for good relations between the Moros and the Americans on Mindanao. Moreover, it is highly doubtful that Filipino leadership alone could have successfully dealt with the Moros.

The Americans, who by chance and not design, were the senior leaders of the Mindanao guerrilla organization could act as honest brokers with the Moros on behalf of the Christian Filipinos. And as chance would have it, Fertig's task was no less difficult than Pershing's.

With the surrender of the USAFFE forces and the dissolution of the Constabulary, whose primary mission on Mindanao had been to contain the Moros, the slaughter began in earnest. The Moros shot or hacked "every man, woman and child they could find," leaving countless hundreds dead, crippled or limbless. In retaliation, Lowland Christian Filipinos plowed their fields with bolos at their belts, formed their own vigilante groups and wiped out entire Moro villages. One of the more infamous Christian leaders, Froilan Matas, a one-time US sailor, led a band which killed some three hundred men, women and children. Christians buried Moros alive, and Moros returned the favor by burying Christians alive in privies.

Juramentado

The animosity between the Moros and Christians was deeply rooted in Mindanao history, and it was something of a cultural imperative. Moros were greatly feared, As it was said, "there never was a Moro who was afraid to die. Death on the field of battle is a privilege and they guard their privileges jealously."[4] The importance of that privilege could be found in the religious rites peculiar to the Islamic Filipinos. The *juramentado* was an elaborately planned ceremony which had connotations of a jihad, or Holy War. But the Moros were not strictly orthodox Mohammedans, and the *juramentado* had no foundation in the Koran.

In this strangely ceremonial kind of suicide-by-killing, the Moro, after careful preparations of his body and clothing and a blessing from his religious leader, would in a frenzy set upon as many infidels (Christians) as he could before finally being felled himself. According to the Moro belief, this death would bring the Moro warrior a martyr's crown, and he would ascend on a white horse to a land of crystal streams and fruit laden trees.

There his passions would be tended to and gratified by dark-eyed *houris*, and his Mora wives would be restored to their virgin youth to contend for his attentions. Each Christian that he had killed would be his slave to eternity in this paradise. As it turned out, the Christians killed

---

[4] Hurley, *Swish of the Kris*, 10.

were borne no particular animosity, for they were victims only by happen-stance. They were tickets to an end for a man who sought death as a blessing.

The *amok*, a desperate, hysterically mad killing, was a distant cousin to the berserker of the Vikings. The difference was that the *amok* was not a killing in battle but could be a frenzied killing of Christian women and children in the village marketplace. It had no accepted religious connotation. For their part, the Japanese did not concern themselves to a large degree with the feuds between the Moro and Christian villages. Albeit, the Japanese themselves fit the definition of infidels, and so they saw fit to curtail the juramentado.

In response, the sultans and datus were called together and told that the punishment for juramentado would be death to the children, parents and grandparents of the Moro who killed. The juramentados diminished, but whether it was due to Japanese or American influence is anybody's guess, although it is probably a combination of both. Not all who were witness to the Moro warriors in action testified to their courage. Fertig did not think that the Moros relished a "stand-up, face-to-face fight," at least not in the Western concept of combat, but he did believe that once cornered, the Moro would "fall somewhat joyously upon his foe, shouting and shrieking, insensible of his wounds."[5]

Some found the Moros to be "very cowardly" and that they would fight only from an "overwhelming advantage."[6] However, there is a difference between courage and stupidity, as Pershing himself observed:

---

[5] *They Fought Alone*, 22.
[6] Edward Haggerty, *Guerrilla Padre in Mindanao* (New York: Longmans, Green, and Company, 1946), 201.

No one who is familiar with Moro character is at all surprised at any turn affairs may take. The more foolish and asinine the thing is, the more likely a Moro is to do it. They make a sudden resolve to die and try to get as many people to die with them as they can. If they are given some time to think this over it is often possible to bring them out from under the influence of the spell.

One example of this is a Moro wedding reception where an insult was exchanged, the krises were unsheathed, and minutes later after the melee had ended forty of the family and guests lay dead.[7]

Fertig had good information on the Moros, and the counsel that Edward Kuder gave to him before Kuder was evacuated to the United States must have been invaluable. Fertig never had any doubt concerning the motives of the Moros. They had little sense of patriotism as the Americans would have understood it, and they had no loyalty whatsoever to the Philippine Commonwealth. Their loyalties were unpredictable. While some actually felt loyalty to the United States, others were ardent collaborators with the Japanese. Most sought separatism as the ultimate goal. And so, for the most part, Moros were tenaciously independent and opposed any outside domination.

Once this was understood, Fertig was able to take a practical approach to his dealings with the Moro chieftains. Who offered what benefits? Whose currency was more valuable? Who pays spies more and which markets for rice were nearer? These questions could be negotiated objectively. The question of most concern to the datus in the long-run, of course, was would the

---

[7] Smythe, Guerrilla Warrior, 169-170.

United States return to the Philippines, and if so, when? As the answer to that question became more and more apparent, the dealing with the Moros did become easier.

Many of the older datus were wily in the ways of politics with the Americans. A chieftain would produce a bamboo tube and withdraw a personal letter to him signed by "Lieutenant Colonel John J. Pershing" or "General Arthur MacArthur." These letters were something of an "anting-anting," a magic charm.[8] As the Moros knew, such "magic charms" could be effective enough when dealing with Americans. Perhaps this is best demonstrated by Sultan Mohammed Jenail Abir. As the spiritual ruler of the Sulu Moros, a half-century before, Abir had fought fiercely against the American soldiers and had surrendered to Captain John J. Pershing. The Sultan kept the American flag flying throughout the Japanese occupation, and for this he was paid a personal visit by Douglas MacArthur.[9]

Securing the loyalty of such prominent Moros was one of Fertig's chief concerns. And by using various stratagems, most tailored to meet the peculiar situation of each datu, Fertig was able to extract promises of support or assurances of neutrality from the Moro chieftains. To seal the deal, Fertig is reported to have made the following threat to them:

> We want the Moros on our side, because we have the same cause – to drive out the invader from our land. Americans have always kept their promises to the Moros, and will continue to do so. You play fair with us, and we will play fair with you; but if you don't, if you attack Christians, loot, wound and slay, we will wipe you out, and if we can't, then the American

---

[8] *They Fought Alone*, 59-60.
[9] *Return to the Philippines*, 203.

Marines will come to our aid and wipe you out to the last man.[10]

This had the desired effect, and the datus gave their oaths to Fertig that they would refrain from stealing until the victory over the Japanese was a reality. Any Moro caught breaking the oath would be executed by the Moros themselves, irrespective of the magnitude of the crime.

Adding a further degree of control, Fertig was even able to place Americans in command of Moro units. The Americans had learned from their early experiments with the Philippine Constabulary that the Moros tended to lionize their leaders. If a leader possessed the leadership qualities of courage and fairness, the Moros would follow him loyally through any hardship. However, the Moros would never serve under a Christian Filipino, and Fertig simply did not trust the Moro leaders to carry out his orders without some influence or presence from him.[11] He remembered too well how General Fort's Moros had deserted him, and he wanted no resurgence of the bloodletting between the Moros and the Christians.

One of Fertig's decisive achievements was to create Moro Militias. One which operated out of the Cotabato region, the Maranao Militia Force, was given to Hedges and fell under the operational control of the 108th Division.

Fertig occupied much of his time by haggling with Moro leaders over agreements for construction of airfields, contracts for the delivery of rice, and negotiations to engage the Japanese, among other

---

[10] *Guerrilla Padre*, 201.
[11] *They Fought Alone*, 151.

things. As an Army surgeon at Camp Vicars had once put it:

> A conference with Moros is a matter of hours, not moments. I cannot conceive anything requiring more tact, patience and courage than is requisite to deal properly with the Moros of Mindanao, and to extract the truth from people who pride themselves on their ability to lie and deceive.[12]

Among the Moro datus with whom Fertig negotiated support was Busran Kalaw who commanded the Fighting Bolo Battalion. Kalaw was the pre-war mayor of Mumungan in Lanao. "His xenophobia was benign. Kalaw was merely anti everyone who was not a Maranao Moro." He was celebrated for his famous letter to the Japanese commander in Lanao, Major Hiramatsu. Hiramatsu had ordered Kalaw to report to Japanese headquarters.

Kalaw wrote back, concluding: "Stop writing and try hiking to attack us so that we will have some more war trophies. Your friend and enemy, Busran Kalaw." Angered, Hiramatsu sent a sizable punitive expedition after Kalaw. It was destroyed to the man, ambushed and trapped in a swamp.[13] Datus Lacub and Dimalaung of Basak pulled the same stunt with Captain Taka Ichi, commander at Dansalan. They destroyed his force of 125 soldiers.

To further solidify his accord with the Moros, Fertig secured the loyalty of the legendary bandit lord Datu Pino. He did this by paying him 20 centavos and one bullet for each set of Japanese ears Pino brought him. By 1945 he was delivering ears by the jar. Pino swore his

---

[12] *Guerrilla Warrior*, 81-82.
[13] *Never Say Die*, 187-188.

loyalty to Fertig and eternal enmity to Morgan. Until then he had been earning his keep by selling carabao bones to the Japanese represented as the remains of Japanese soldiers so that they could receive a proper burial according to Japanese custom.[14]

Some Moros, however, were non-committal. Sultan sa Ramain was neutral, but he agreed with Fertig that he would sell rice to the Japanese and then tell Fertig when the Japanese trucks would pick up the rice. He got paid, the Japanese got their rice, and the guerrillas got both the Japanese and the rice.[15]

Not all Moros supported the guerrillas or were neutral, of course. Datus Pain and Sinsuat Balabaran openly supported the Japanese. Many others played both ends against the middle. Many of the difficulties with the Moros arose simply because they refused to work with or serve under one another. Family rank and social standing were more important than the military titles.

Though unusual, there were Moro leaders who worked well with Christians on their staffs, or at least there was no open violence. Datu Ogtog Matalam, a powerful leader in Cotabato, arranged for sacraments for his Christian guerrillas. One of the more picturesque of the minor guerrilla leaders was Captain Hamid. Hamid was a Moro who wore a crucifix, was married to a Christian Filipina, and commanded a battalion of Christian guerrillas.[16]

---

[14] *Return to the Philippines*, 158. The trade in enemy trophies went both ways. The Japanese paid the Moros 2,000 pesos for the head of an American or Australian. Hal Richardson, One-Man War: The Jock McLaren Story, 1957, 64; Keats, They Fought Alone, 284-285.

[15] *They Fought Alone*, 306.

[16] *Guerrilla Padre*, 136, 138.

Thus, against great odds, certainly against any historical probabilities that the plan would work, the Mindanao resistance movement was pulled together. Post-war writers wrote colorful assessments like "never before in the history of our country did the Moro and Christian Filipinos understand and cooperate with each other more closely than during the resistance movement in Mindanao."[17]

Fertig's truce with the Moro effectively neutralized the Moro threat in the Misamis area. The loose alliance was not long-lasting, but it lasted long enough to keep the guerrilla organization unified and the resistance movement alive. In the last analysis, one of Fertig's greatest achievements that had a profound impact on the course of the resistance was his ability to weld such disparate groups together under one command. The momentum to resist the Japanese was smoldering on Mindanao. With this Moro accelerant, Fertig now had all the ingredients he needed to fan the scattered brush fires into a raging inferno – a fire in the jungle.

---

[17] See Office of Strategic Services, Guerrilla Resistance in the Philippines, July 21, 1944, 303.

Chapter Six
MacArthur's Unconventional War

"All this was done in a saga of blood and loyalty that I can never forget." – MacArthur

Prior to the fall of the Philippines, General MacArthur initiated plans to establish a guerrilla structure in the islands. However, the Japanese invasion cut these plans short. As events unfolded in the Philippines, the Joint Chiefs of Staff divided the Pacific theatre into the Pacific Ocean Areas (POA), under Admiral Chester Nimitz and the Southwest Pacific Area (SWPA), under MacArthur.

In the weeks that followed, SWPA received radio messages from the Philippines. Praeger, on Northern Luzon, had some 5,000 guerrilla fighters. Likewise, Colonels Moses and Noble could muster approximately 6,000 men on Luzon. While the guerrillas struggled to survive and build their organizations, their constant appeals for help had been reaching MacArthur's headquarters some 3,500 miles to the south in Australia.

A keen student of military history, MacArthur was well aware of how guerrillas can erode the ability of conventional forces to fight. His own father experienced the frustrations of counterguerrilla warfare while leading American troops in the Philippines during the Philippine Insurrection. And so, developing an unconventional warfare campaign was an important part of MacArthur's plan for the liberation of the Philippines. And while others discounted it, MacArthur believed the intelligence provided by guerrillas was a valuable asset and prerequisite to a retaking of the Philippines.

Having this guerrilla capability, and countless contacts, together with an auxiliary force in place, MacArthur and his staff, were able to galvanize a resistance movement among the Filipino people. MacArthur began by creating the Allied Intelligence Bureau (AIB) in June 1942. Consisting of US, British, Australian, and Dutch intelligence and special operations personnel, it was formed to coordinate guerrilla, subversion, and propaganda organizations.

The AIB conducted everything from coastal reconnaissance (what would be called coastwatchers), and commando raids, to the smuggling in of thousands of tons of weapons and supplies via submarine from remote bases in Australia. It would coordinate the existing Allied propaganda and guerrilla organizations. The role of the AIB was to obtain information about the enemy, to weaken the enemy by sabotage and destruction of morale and to lend aid and assistance to local effort to the same end in enemy territories.

As the guerrilla movement grew in size, MacArthur assigned Colonel Courtney Whitney to manage the effort. MacArthur's directives for the new effort was as follows:

> His objectives were the formation of a battle detachment in every important Filipino area, alerted to strike against the enemy's rear as our battle lines advanced; to secure fields adjacent to military objectives into which our airmen might drop with assurances of immediate rescue and protection; to arouse the militant loyalty of a whole people by forming resolute armed centers of resistance around which they could rally; to establish a vast network of agents numbering into the thousands to provide precise, accurate, and detailed information on major enemy moves and installations; to create a vast network of radio positions extending into every center of enemy activity and concentration through the islands; to build on every major island of the Philippines a

completely equipped and staffed weather observatory to flash to my headquarters full weather data morning, noon, and night of everyday; to implement an air-warning system affording visual observation of the air over every square foot of Philippine soil to give immediate warning of aircraft or naval movement to our submarines on patrol in Philippine waters; to apply close interior vigilance so as to secure for our military use enemy documents of value; and to exploit any other aids to our military operations that might arise.[1]

All of which was quite a task, and as MacArthur adds, it was accomplished "in a saga of blood and loyalty that I can never forget." The following map details the major resistance movements (See figure 6-1).

In these early stages of development, the Office of Strategic Services (OSS), which was created to conduct operations just of this sort, repeatedly offered to send agents to assist MacArthur. However, MacArthur consistently refused the offers, and Willoughby assured him that they were not needed. As D. Clayton James observes, MacArthur was "not about to have Allied personnel in his theater who were not under his control, as would have been the case with the OSS."[2]

Arguably, MacArthur already had several organizations available to him in Melbourne. He had the British SOE to conduct sabotage and espionage, he had the Netherlands Indies Forces Intelligence Service (NEFIS), and he had the Australian propaganda units and their coastwatcher network. For MacArthur the AIB would be his counterpart to the OSS, although it would have far fewer financial resources available to it.

---

[1] Douglas MacArthur, *Reminiscences* (New York: Ishi Press, 2010), 205.
[2] James, *Years of MacArthur*, II, 510-511.

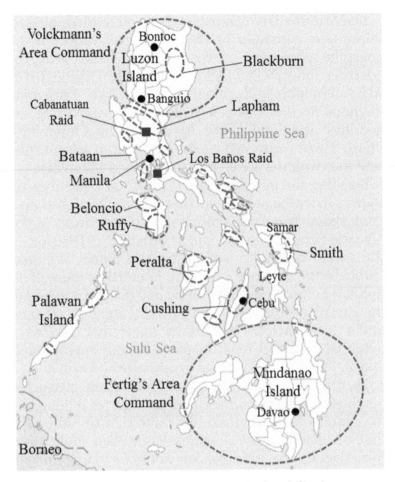

Figure 6-1. Resistance Movement in the Philippines.

MacArthur did his best to keep the OSS out of his theater, and the small amount of work done by the OSS on the Philippines reflects this. There are virtually no OSS reports on the Philippine resistance movement, and the information the OSS did put together was brief, summary in nature, and came from interviews and not from their own agents in the Philippines.

MacArthur's UW campaign unfolded in three phases: Phase One consisted of the initial exploration of the guerrilla movement by the AIB. In Phase Two, the guerrilla movement would be developed through the AIB's Philippine Regional Section. The Philippine Regional Section was solely devoted to supporting the guerillas and intelligence agents in the Philippines. Phase Three composed of the merging of all guerrilla activities with the actual invasion of the Philippines.[3]

Boosting the morale of MacArthur, on December 12, 1942, a listening post in northern Australia picked up a faint signal from the Philippines. It was from Major Macario Peralta who had fought with the 61st Division on the Island of Panay. According to Colonel Whitney, "Probably no message ever gave MacArthur more of an uplift."[4] Peralta reported that he had taken command of the guerrilla forces on Panay and had some 8,000 men under his control.[5] MacArthur knew that he had a potential powerful force in place behind enemy lines, and all that was needed was to organize and arm it.

Then in early December 1942, Macarthur instructed his chief intelligence officer, Colonel Charles Willoughby to organize "a commando-like force that would be the first to penetrate the Japanese-controlled Philippines from the outside."[6] Its mission would be to link up with these disparate bands of guerrillas and get an on the ground assessment of the overall capability. To this end,

---

[3] Douglas MacArthur, *Reports of General MacArthur: The Campaigns of MacArthur in the Pacific, vol. 1* (Washington, DC: US Army Center of Military History, 1994), 298.

[4] Courtney Whitney, *MacArthur: His Rendezvous with History* (New York: Alfred A. Kropf, 1956), 128.

[5] William B. Breuer, *MacArthur's Undercover War: Spies, Saboteurs, Guerrillas, and Secret Missions* (New York: John Wiley & Sons, 1995), 47.

[6] Ibid., 48.

US Navy Commander Charles "Chick" Parsons was called upon to lead an organization known as Spy Squadron (SPYRON). As a joint effort, SPYRON fell under the US Navy's 7th Fleet while SWPA's intelligence section administered the program through the AIB.

The plan for SPYRON was as follows: Parsons and his group would clandestinely make contact with these various guerrilla bands, vet the leaders, provide supplies and the means of command and control via radio networks. MacArthur believed the guerilla movement was a critical element to a victory the Philippines.[7] These tasks would therefore be fundamentally important to the success of the resistance movement.

---

[7] Ibid.

Chapter Seven
Fertig's Fiefdom

"Even after a defeat, there is always the possibility that a turn of fortune can be brought about by developing new sources of internal strength or through the natural decimation all offensives suffer in the long run or by means of help from abroad. There will always be time enough to die." — Clausewitz

The early bandit groups had given the guerrillas a poor image among the civilian populace. These groups expropriated property and food from the civilians, and the new word coined for taking something without payment was "USAFFED." For example, as in the expression: "the guerrilla USAFFED my carabao." In Visayan the new word was "Tulisaffe," which meant USAFFE thief.

To avoid such appellations, Fertig, very early in the establishment of his headquarters, abandoned forced requisitions on any kind and relied upon a system where the guerrilla organization would pay its own way. Many guerrillas paid their own way with personal promissory notes during the occupation. For example, Lieutenant Colonel McClish's 110th Division in Agusan had a guerrilla dance band which raised money by playing for weddings and fiestas. They charged 50 centavos for admission to their nightly dances. The band was called the "Best Band West of Ford Island, Pearl Harbor."

With all the positive steps behind him, Fertig knew the guerrillas would need to be paid. He therefore issued scrip to restart Mindanao's economy. In fact, if he didn't, he would appear to be a liar to, as he had promised the Filipinos and Moros that they would be paid if they joined his army.

Fertig's little war had drawn desperados from all over the island. One such man was Fertig's old friend Sam Wilson. Wilson had been sent from Corregidor to Mindanao in February 1942 on a mission. Wilson came stumbling out of the hills, half-starved and still accompanied by a porter carrying his mattress.

In civilian life Wilson was the owner of the Wilson Building in Manila. A millionaire, he had made his fortune speculating in mining stocks and real estate. A middle-aged man who knew nothing of the jungle, he had been commissioned a lieutenant in the Navy just before Manila fell. His wife and boys were interned by the Japanese in Santo Tomas on Luzon, and he feared for their lives should the Japanese learn of his guerrilla activities.

Wilson was the only man Fertig pleaded with to join the resistance. And when he did, he ran the money printing press and handled the guerrilla finances. He became a comptroller and secretary of the treasury all in one. In fact, Wilson's currency notes provided a uniform currency for the entire island (See figure 7-1).

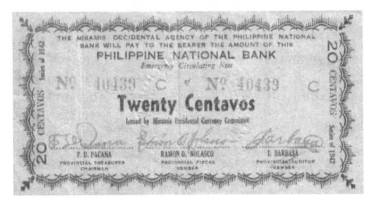

Figure 7-1. Misamis Occidental provincial notes of 1942.

Wilson's mint was always ready and packed to go on a moment's notice, with all receipts signed and all accounts current. In all, Wilson's mint was moved a total of six times during the war. On occasion, when a Japanese assault forced Fertig to move his headquarters, Sam Wilson's clandestine mint was back in working order the next day.

Despite its success, MacArthur banned the printing, referring to it as military scrip, or promissory notes. This order made the printing of the emergency money an outright disobedience. Again, Fertig was faced with another dilemma. If he ditched the mint, he would appear to be a liar to his own people, as he had promised the Filipinos and Moros that the emergency notes would be redeemed at face value after the war was over. He neatly side-stepped the issue by reasoning that he wasn't printing military scrip, but real money; money which sustained the guerrilla economy and kept the war going. He told Sam Wilson: "Hell, if those chair-bound commandos in Australia want to come here and eat cold rice and dodge Japs with me while we debate whether its money or scrip, I'll gladly arrange it."

As it turned out, the situation remedied itself. Early in 1943, exiled President Quezon authorized the creation of the Mindanao Emergency Currency Board which had the authority to issue its own monetary notes for use as the medium of exchange among the guerrillas. With wooden plates and paper supplied by submarine, the guerrillas printed their own currency. All bills were carefully numbered and recorded in two separate ledgers, and they were to be redeemable by the US

Government at war's end. The penalty for counterfeiting was "severe."[1]

Fertig soon became concerned in these early weeks that he was not recruiting the kind of people needed to sustain the resistance movement. He was enlisting "generally masterless men, adventurous youths, men who wished to get away from their wives, and the kind of men who seemed to have nothing better to do."[2] Moreover, vital to everything, Fertig saw that the civil government needed to be reinstated. It was a necessity. Fertig relied upon funds gained through taxation by the civil government and the creation of money-making projects run by the guerrillas.

The establishment of civil government in Mindanao under guerrilla auspices did much to further strengthen the resistance movement. In October and November 1942, Fertig ordered the screening of all former government officials for the purpose of filling positions in the newly established civil government of the Free Philippines. Using as his authority, Fertig ordered the pre-war Mindanao government officials back to work in their former positions.

The Province of Free Lanao was officially established on December 1, 1942 under Governor Marcelo T. Paiso at Causwagan near Kolambugan. Paso commenced the reorganization of the municipal governments in the unoccupied territories. The Free Lanao government was so successful that Fertig sought to achieve the same success in other provinces using the same procedures. In the areas where the Japanese were strong and the

---

[1] *They Fought Alone*, 152, 205; Hawkins, Never Say Die, pp. 168-169; Agoncillo, Fateful Years, pp. 746-747; SCAP, Reports, I, p. 309.
[2] Baclagon, Philippine Campaigns, p. 282; Ltr Fertig to Casey; HQ, Philippines Command, "Recognition Program," p. 41.

establishment of a free civil government deemed impractical, Fertig established martial law under the 10th Military District.

This was done in Bukidnon and a portion of Misamis Oriental. Interestingly, in some areas government officials had to serve two masters but Fertig considered the gains were worth the risk. The question of who served the resistance movement would be raised at a later time, and everyone knew that. During Parson's first visit to Fertig, the issue of civil government was raised. MacArthur's staff was concerned that such initiative would interfere with the Philippine government in exile.[3]

It therefore came as no surprise then, when President Sergio Osmeña and General MacArthur announced the following policy for the reestablishment of the Philippine government on liberation of occupied areas by the American forces:

> So far as possible, provincial and municipal officers last serving under the authority of recognized guerrilla leaders or in recognized free governments will occupy positions of equal or better rank as temporary officials.

Fertig ran the guerrilla organization like a business, and the form some of its functions took reflected this phenomenon.

Fertig was fortunate to be on Mindanao, however, for the problems of food supply were less acute here than they were on the other islands, and many of the better food producing areas were in guerrilla-controlled areas. Fertig's problem was one of distribution, because for food to be delivered to the guerrillas and civilians in the

---

[3] Kent Holmes, *Wendell Fertig and His Guerrilla Forces in the Philippines* (Jefferson, NC: McFarland, 2010), 46.

areas which were barren of agriculture the supply trains had to traverse miles of mountains and pass through Japanese-held territory.

Fertig created the Food Supply Administration and the Trading Post Administration to monitor the growing of food, see to the payment of the workers, and coordinate the movement and security of the supply columns as they moved off through the mountains.

The problem was not an easy one, for a cargadore would often consume more than he could carry for a three-week journey through the mountains. And, of course, he still had to return. Caravans of 100 or more heavily-laden carabaos were not uncommon. Many of the cargadores hired were Atas, nomadic aborigines who hunted and fished, constantly staying just ahead of civilization and areas exhausted of game.

The Atas stood four feet tall and weighed 90 pounds, but they could carry a 75-pound load with little effort, and they navigated in the jungle quite easily. They made ideal cargo bearers or they could hunt their own meals as well. A Women's Auxiliary Service was formed by Josefa Capistrano, a Chinese Mestiza. Her husband, Nick Capistrano, a Mestizo and wealthy engineer, ran the labor groups, cargadores, oversaw the mechanical equipment, and scheduled the collection and distribution of food to the guerrilla force. After the war there were problems with compensation for those people who had assisted the guerrillas before the GHQ, SWPA official recognition date of February 13, 1943. Fertig had, of course, been issuing IOU's on the US government before this date.

Fertig established a Civilian Relief Administration which decentralized its powers to the Directors of the Provincial and Municipal Civilian Relief Committees.

The Provincial Relief Committee helped the families of the guerrilla fighters. In addition, each province had its own Director of Civil Affairs who dealt with any unusual problems caused by guerrilla units in the province. Hoarding and black-market profiteering were not widespread problems, but they could have a major impact on a small village. The penalty for hoarding was a jail sentence. The Provincial Emergency Control Boards created by Fertig fixed the prices on prime commodities, but concentrated primarily on the prices of corn and rice because the prices of all other goods appeared to fluctuate in proportion to the price on these two commodities.

In the end, the issue of pay was a difficult one. Through Executive Orders Numbered 21 and 22 exiled President Osmeña confirmed that the service of the guerrillas would be rewarded with recognition and that recognition would bring them pay, allowances and benefits of soldiers belonging to the Philippine Army, which was the same as the pay for the American soldier because the Philippine Army was a part of the US Armed Forces. GHQ, SWPA in advertently officially confirmed this in Circular Number 100 on November 17, 1943. This was never the intent of either government, for the US pay scales would be highly inflationary in the post-war Philippine economy and would put the common soldier in the highest pay brackets of the government.

To avoid profiteering among the guerrillas their leaders gave them only partial pay: 10 pesos per month for privates and up to 150 pesos per month for field grade officers. In any case, there was little to buy in the guerrilla-controlled areas, and the guerrilla currency was no good in the Japanese-occupied areas, so the guerrillas called their script *"tinghoy,"* meaning counterfeit or useless.

MacArthur through Osmeña authorized 50 pesos pay per month for each guerrilla to be claimed on the US Government after the war, but the US Government finally allowed only eight pesos per month when the final decisions were made after the cessation of hostilities. The effects of this were several. Colonel Fertig's official troop lists for claims purposes were altered after the war by Philippine Army and Philippine Government officials in Manila. Names were deleted, and others were added, usually the names of relatives of the officials. The great promise of the United States to its guerrilla forces and the people who supported them was broken. The hard-won scraps of paper did not mean what they had been promised to mean.[4]

As soon as the war was over the question of which guerrilla issues would be redeemed by the Philippine and American governments arose. A period of four months was allowed for the exchange of guerrilla money for post-war Philippine Commonwealth notes.

---

[4] Ingham, Rendezvous, p. 169; Smith, "Status of Members," pp. 40-41; HQ, Philippines Comand, Recognition Program" p. 98, See especially Appendix 13; Friend, Between Two Empires, p. 255; Keats, They Fought Alone, p. 413. The benefits available to the guerrillas after the war through the U.S. Veterans Administration are contained in Dormal, War in Panay, 276-178.

Chapter Eight
10ᵗʰ *Military District*

"Like smoldering embers, it consumes the basic foundations of the enemy forces. Since it needs time to be effective, a state tension will develop while the two elements interact. This tension will either gradually relax, if the insurgency is suppressed in some places and slowly burns itself out in others, or else it will build up to a crisis: a general conflagration close in on the enemy, driving him out of the country before he is faced with total destruction." – Clausewitz

After establishing the nucleus of his guerrilla organization, Fertig next confronted with the problem of communicating with a higher headquarters – MacArthur. On December 4, 1942, at their own instigation and over the skepticism voiced by Fertig, Captains J. A. Hamner and Charles Smith left Mindanao by small sailboat with a Moro crew for Australia. Both Smith and Hamner had been mining engineers before the war, both knew Fertig, both had been at Jacob Deisher's mountain camp when it was attacked by Moros. Their odyssey was successful, and all the more so because neither was an accomplished blue water sailor.

In the meantime, Fertig had initiated attempts to communicate with Australia by radio. The story of how he made radio contact with MacArthur is a fascinating one. A member of Fertig's headquarters, Gerardo Almendres, a high school boy, had some books he received before the war from the International Correspondence School, Scranton, Pennsylvania. With no experience whatsoever, Almendres took bits and pieces from old radio receivers and sound equipment parts from an old motion picture projector that had been buried in a swamp and tried to duplicate the diagrams in his books.

The resulting configuration covered four walls of a nipa shack (grass hut). Robert Ball, an Air Corps radio operator, and Roy Bell, a school teacher and ham radio operator from the island of Negros, solved the problem of the aerial and the crystal, which Almendres' radio lacked, by using wire coiled erratically around a joint of bamboo. The radio was tried every day, taken completely apart and tried again the next, looking for the right combination of parts. The signal transmitted did not keep to one frequency but slid across the kilocycle band interrupting traffic on all frequencies.

When the signal was first picked up by the Station KFS (Signal Intelligence Division station of the Western Defense Command) at Half Moon Bay, San Francisco, the Navy signalman thought the Japanese were trying to jam the radio waves. In fact, for weeks they thought Fertig was anything other than a Japanese counterintelligence agent posing as Fertig. FBI and Naval Intelligence agents even visited Fertig's wife in Golden, Colorado and reviewed his military files. A break came when KFS called using "MSF" as the call letters. This was a code Captain Smith had taken to Australia. It stood for "Mindanao Smith Fertig."

Now the War Department and General Headquarters South West Pacific Area (GHQ, SWPA) were involved, and Smith was there to vouch for Fertig's existence. Still, it was not until February 14, 1943, that KFS was satisfied with Fertig's authenticity and radioed procedural instructions for contacting Station KAZ, MacArthur's headquarters in Australia.

Contact was formally established with KAZ on February 20, 1943.[1] Fertig's first report to MacArthur read: "Have strong force with complete civilian

---

[1] Mindanao was the fourth station in the Philippines to do so.

support...large number of enemy motor vehicles and bridges have been destroyed. Many telephone poles have been torn down, food dumps burned, and considerable enemy arms and ammunition captured. Thousands young Filipinos eager to join when arms available. Ready and eager to engage the enemy on your orders."[2] MacArthur's first message read as follows:

> FERTIG XXX YOU ARE NAMED GUERRILLA CHIEF XXX YOUR MEN ARE NOT DESERTERS BUT FIGHTERS XXX IN SOME WAY I WILL GET AID TO YOU XXX FOR THE FUTURE I REITERATE MY PLEDGE XXX I SHALL RETURN XXX MACARTHUR

This message brought relief to the USAFFE soldiers who were still unsure of their status under military law, and it held out the prospect for help. A following message appointing Fertig to command was still more specific:[3]

> LTCOL W W FERTIG IS DESIGNATED TO COMMAND THE TENTH MILITARY DISTRICT (ISLANDS OF MINDANAO AND SULU) XXX HE WILL PERFECT INTELLIGENCE NET COVERING NINTH MILITARY DISTRICT (SAMAR-LEYTE) XXX NO OFFICER OF RANK OF GENERAL WILL BE DESIGNATED AT PRESENT XXX

For now, his internal problems at an end, Fertig was free to concentrate on consolidating his force. And so, by perseverance and diplomacy, Fertig gradually won the respect of the other guerrilla leaders, and built up a fairly cohesive guerrilla organization.

Thousands of miles to the south, from Australia, MacArthur and his staff organized some 270 guerrilla

---

[2] MacArthur, *Reminiscences*, p. 204.
[3] Steinberg, Return to the Philippines, p. 24.

units under one command, GHQ, SWPA. This was no easy task. MacArthur sent a message to Fertig on February 13, 1943 which specified several things. He limited Fertig's command to the 10th Military District which included the following: Mindanao, Zamboanga, the four islands in the Province Surigao del Norte, and Basilan Island off the tip of Zamboanga. This message declared MacArthur's intention to develop the guerrilla territorial command area based on prewar military districts.

All district commanders would function under the control of GHQ, SWPA. MacArthur had not forgotten that a single commander for all of the islands if captured could surrender all of the forces. But more important, he wanted no contest among the district commanders for overall command of all the guerrilla forces. In the February 13th message he spelled this out very clearly: "It is directed that there will be maximum co-operation, mutual support and the avoidance of friction between commanding officers of military districts operating toward a common cause.

With good reason MacArthur had emphasized the need to eliminate competition. He knew that strong personalities would rise to leadership and would begin jockeying for power. There is good evidence that MacArthur's message was, in good part, directed at Fertig. Earlier, Fertig held aspirations for control of the Visayas, an area under the control of Colonel Peralta.

Before Fertig established communication with MacArthur, Peralta radioed MacArthur and complained that "certain officers, including one Wendell Fertig, were trying to usurp his command" and had requested that MacArthur recognize him, Peralta, as the sole commander of the Philippine forces. That bad blood existed between the two is suggested by Keats' description of Fertig's reaction. When Fertig learned of Peralta's message, "if Peralta had been in Fertig's office at the moment, Fertig would have drawn his pistol and shot Peralta down where he stood."[4]

The issue of GHQ, SWPA recognition was an extremely important one to the guerrilla leaders, and Fertig and Peralta were fortunate to have received recognition early. It enabled them to quickly consolidate their forces with the respectability and authority derived from their appointments. Early recognition was advantageous to GHQ also because it could then pursue within that command what came to be known as the "lie low" policy. What MacArthur needed most from the guerrillas in the Philippines was intelligence. As intelligence gatherers, the guerrillas could give the American forces a tremendous advantage in planning military operations.

In return for recognition, a guerrilla leader would receive aid, but only upon his meeting four conditions. The guerrillas were to remain united under one command within the military district; the guerrillas must not usurp but must support the local civil government; combat activity against the Japanese was to be limited to protection of guerrilla facilities; and the leader was to faithfully carry out his responsibility to strengthen his organization and gather intelligence.[5]

In order to guide the various guerrilla leaders in the prosecution of their operations and to make maximum use of their services in the war against Japan:

---

[4] *They Fought Alone*, 184-185.
[5] Dissette, Guerrilla Submarines, 31.

Guerrilla groups were advised to assist in maintaining civil order so that they might receive reciprocal popular support. They were also cautioned to refrain from open and aggressive warfare against Japanese troops lest they bring reprisals on the people out of all proportion to the results achieved. The collection, coordination, and transmission of useful intelligence were stressed as the most important, immediate contributions the guerrillas could make to the Allied cause until the actual invasion of the islands was begun. Before that time, all military operations were to be limited to strategic harassment, sabotage, and ambush.[6]

The most that was therefore expected of the guerrillas "operationally" was sabotage of the Japanese lines of communication.

Foremost in the minds of MacArthur's staff was the fact that the Japanese would vent their wrath on the civilian population if the guerrillas persisted in attacks. Moreover, the guerrillas, given their level of training, could not hope to cope with a Japanese attack in force. If this occurred, guerrilla strength would be reduced and their ability to collect intelligence would diminish. In light of this, MacArthur issued a "lie low" policy to reduce the likelihood of reprisals.

Guerrilla HQ

Crucial to the development of the Mindanao guerrilla organization was the location in the early months of the guerrilla headquarters and the relationship established with the civilians. The provincial government of Mindanao continued to function for the seven months after the fall of Manila. This factor coupled with the Japanese garrisoning plan provided the "incubator" in which the guerrillas could gain their strength in infancy.

---

[6] The Campaigns of MacArthur in the Pacific, Volume 1, Chapter X, 304.

After the surrender in May 1942, the Japanese concentrated their garrison forces in Davao, Lake Lanao and the Cagayan-Illigan-Kolambugan area with garrisons in twelve towns. They further sent a small guard detachment to Misamis Occidental Province, but it had been withdrawn. Misamis Occidental was of no strategic value to the Japanese, and neither had it been viewed as such by the American island defenders. But USAFFE soldiers had drifted into the area to avoid the Japanese, and this was the province that Luis Morgan and William Tait had so easily secured under their control. In eastern Misamis Occidental was Misamis City, a small town situated on Panguil Bay. Lying just east of Mount Malindang, this town was the original site of the guerrilla headquarters.

Misamis had been important as a port and business center for the area's agricultural concerns. It was a trade center and shipping point for the Kolambugan Lumber Company, and there was an old Spanish fort still standing in the town. Fertig made the fort his headquarters and flew the American and Filipino flags at equal height from the fort. The province had 250,000 people living in it, and the civil government had virtually continued to function after the surrender because the Japanese had not bothered with the area.

For those guerrillas who came down out of the mountains, disease ridden and half-starved, the business as usual attitude in Misamis City must have made it seem as though they had gone through a time warp to a time before the war. Schools were open, priests held mass, shops were open and food was plentiful. There were even guest quarters for couriers and visitors with beds, hot showers, and dinner. Moreover, money was being printed, *banca* boats loaded with trade goods filled the bay, factories were humming, and the lights were on.

Inter-island trade was flourishing and an estimated 50 percent of the manufactured goods in the Visayas were exported among the southern islands. The telephone system worked: soda pop bottles were used for insulators, fencing wire replaced copper wire, and telephone batteries were recharged overnight by being soaked in tuba (coconut beer). The telegraph system was also working, and the trucks from Hedges' old 81st Infantry Division motor pool had been retrieved from their hiding places in the jungle. Fertig was communicating by courier with the islands to the north, and one American family hiding deep in the mountains of southern Negros took hope when told of a "General" Fertig who had been sent by MacArthur to Mindanao to train a guerrilla army.

Anchors Aweigh

To evaluate Fertig's guerrilla organization, MacArthur sent Charles "Chick" Parsons. His mission was to learn the extent of the resistance movement in the Philippines, gauge the ability and trustworthiness of the leaders, and identify those who were capable and willing to accept orders from MacArthur. Those who could meet the standards would receive official recognition, which Parsons had the authority to offer, and the assurance of supplies. Parsons would then integrate these units into an archipelago wide intelligence network.[7]

When Parsons arrived on the *Tambor* in March 1943 he had no expectation that he would encounter anything like Misamis City. What Parson saw went far in

---

7 Wise, *Secret Mission*, 76.

convincing him that Fertig could lead and that the resistance movement had a chance. The crew of the *Tambor* was met by an orchestra dressed in white playing "Anchors Aweigh." One sailor remarked that the submarine's skipper had taken a wrong turn and ended up in Hollywood. Fertig's purpose in all of this was multifold: he wanted to impress SWPA that he had a viable organization. More than that, he wanted the Filipinos on Mindanao and elsewhere to know that American aid was coming.

When met by a guerrilla leader on the beach, Parsons asked how the submarine could be unloaded of the supplies he had brought for the guerrillas. The guerrilla promptly produced a lighter to transport the two tons of stores brought aboard the *Tambor* to the beach. It took 40 minutes to move the shipment: 50,000 rounds of .30 caliber and 20,000 rounds of .45 caliber ammunition; radio equipment and spare parts; medicine, clothing, food, soap, cigarettes, bandages and surgical kits. There was even a can of wheat flour for making communion wafers, and $10,000 in cash.

Fertig was impressed with the amount of supplies brought and knew that their delivery to him would raise the level of respect for his position as fledgling leader of the Mindanao guerrilla movement. To a starving man a piece of bread is a banquet, so it is no surprise that Fertig was appreciative of the effort. And it must have seemed plentiful to those who had to carry it. Still, it was very small, and Fertig knew it – it amounted to only a pound per man in his small force.

When he pressed Parsons on how frequently future deliveries might come, he was told that supplies were coming only because MacArthur was personally interested in the welfare of the Filipino guerrillas. The

War Department was not especially interested in the guerrillas, for it had concerns of much greater magnitude. The guerrillas saw the situation quite differently. It is not an overstatement to say that each action against a Japanese patrol on Mindanao was seen as a great Allied victory by the guerrillas, and for them that was the war in the Pacific.

General Morimoto, the Japanese commander for Mindanao-Sulu belittled this first supply effort in a public communique and said: "The only thing I wish, should anyone find American cigarettes, bring me a package so I may smoke."

The situation at Misamis City was not all rosy of course. The beaches were laced with barbed wire, a continuing sign that the Japanese could come at any time. Indeed, every morning a Japanese reconnaissance plane arrived at exactly eight o'clock, took pictures, and dropped a bomb or two on the old fort. The people would simply leave the fort area at that time every day to run errands elsewhere. Many of the bombs dropped were duds and these became a reliable source of gun powder.

To bolster manpower, Fertig revived the Home Guard. This was an old Spanish system of *voluntarios* which required each man to devote so many days of labor per month to the government in lieu of taxes.

The *voluntarios* were not part of the regular guerrilla force, and they were used to guard trails and roads and to provide early warning of Japanese patrols, which passed through the area about once a month. By this time, Fertig claimed a guerrilla force of some 15,000 with 5,000 rifles. He even established schools for his guerrillas. There was an Officer Candidate School, a commando course, and various other training schools. As the resistance grew, Fertig widened his gaze. By

February 1943, he was able to report to SWPA the possession of two airfields in Bukidnon Province with 2,000 troops available to protect the fields.[8]

Fertig very early on established the tenor for the command's administration by making copies of regulations, troop lists, finance records, orders, and records of guerrilla operations. He saw the obvious need for documentation to support post-war claims on the American and Filipino governments. But he also appeared to believe that the show of administrative activity gave some kind of mystic legitimacy to his command.

The stamp "file copy" somehow created the image of organization and stability. Fertig was apparently convinced that Filipinos are "impressed by the flood of official-seeming documents." As time went on, of course, the few typewriters fell into disrepair, typewriter ribbons shredded, and paper of any kind was at a premium. As head of the administrative staff, Fertig also found himself involved in a multitude of seemingly trivial routine matters.

In some ways, the flurry of administration was much like leaving the lights in Misamis on at night. Like the lights, the file copy was symbolic for it showed that the guerrilla force was there to stay and would not desert the people. In this case substance followed form: the concept was to look and act like an organized force.[9]

Fertig was under no illusions that the location of his headquarters would be permanent. He was equipped for instant mobility, and he kept his maps, codes, sensitive intelligence information, and current business in a briefcase which he could grab and thus leave quickly.

---

[8] *Guerrilla Padre*, 73.
[9] *They Fought Alone*, 107, 115, 134-135, 290.

The briefcase, which was brought by Parsons from Australia, was a trick briefcase which was designed to explode into a magnesium fire upon being opened unless a hidden switch was properly activated. Other records of his command were buried in camouflaged holes in the ground in sealed tin cans.[10]

Overall, the visit by Parsons to Mindanao marked an important turning point in the resistance movement. Fertig knew that Parsons and Smith had been sent to determine if he was competent to command, and he was told outright that he had not done himself any favors at GHQ by promoting himself to brigadier general. The issue of Fertig's competence was probably never in doubt. Parsons knew McClish, Bowler and Hedges personally, and he had confidence in both their abilities, as well as their discernment.

In the last analysis, Parsons concluded that the Mindanao resistance movement was stronger than GHQ had contemplated. He also concluded that Fertig needed small arms, ammunition and radio equipment for intra-island communication. He confirmed Fertig as the commander of the 10th Military District and produced a set of silver eagles to confirm his promotion to colonel. The designated boundaries for the ten military districts matched with the pre-war Philippine Army districts (See figure 8-1).

Pendatun

One of the most important accomplishments of Parsons' stay on Mindanao was the assistance he provided in unifying the Mindanao guerrillas, namely that of bringing Salipada K. Pendatun, a guerrilla leader in the Cotabato Province, into the Fertig fold. Pendatun

---

[10] *American Guerrilla*, 133.

was the number one troublemaker as far as Fertig was concerned. Parsons travelled to Cotabato to meet with Pendatun and persuade him to submit to Fertig's authority.

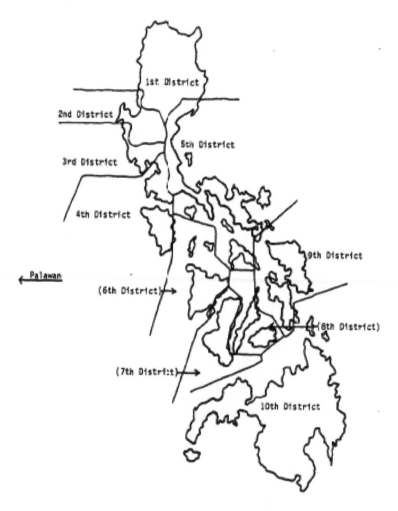

Figure 8-1. MacArthur's Organization of the Philippines.

The rift all began in December 1942, when Pendatun offered Fertig a job on Pendatun's staff. Pendatun was a lawyer and an influential adviser to the Governor of Mindanao before the Japanese invasion. A Cotabato Moro and son of an old line of datus, Pendatun's men acknowledged him as "brigadier general." Moreover, Pendatun had a competent staff, better than Fertig's which included Americans, a former senator, a former governor of Cotabato, the former chief of staff of the Philippine Air Corps, and Major Frank McGee. Fertig thought that Pendatun's group was very effective but found Pendatun "headstrong, brave, glittering" and "overly occupied with the pleasures of women."

Parsons visited Pendatun and travelled with him. He was impressed with his several thousand men and his staff and the manner in which the guerrillas grew their own produce and livestock. In an apocryphal tale, Parsons was with Pendatun when he trapped a Japanese patrol inside a schoolhouse. Unable to reduce the building with small arms, Pendatun resorted to strapping two bombs on a carabao, lighting its tail and pointing the terrified animal at the building.

The method enjoyed some small degree of success, but Parsons was now able to make it clear to Pendatun that the only way he would ever get heavier weapons would be by acknowledging Fertig as the leader of the Mindanao guerrillas and joining MacArthur's team. As it has been recounted, perhaps with some poetic license, at that, Pendatun gave up his dream of commanding General Vachon's former Bukidnon-Cotabato Force, and on the spot removed his gold stars and replaced them with gold oak leaves that Parsons had brought with him.[11]

---

[11] Wise, Secret Mission, p. 104; Ingham, Rendezvous, pp. 84-86.

The transformation was not quite so smooth. In fact, it was not until Fertig was moving his headquarters to Agusan in the fall of 1943 that Pendatun at last capitulated. The example of Pendatun is the most important one for demonstrating the gloved fist used by Fertig to persuade guerrilla leaders to join his command. Pendatun led a powerful organization, and he had the capability to spoil Fertig's efforts on Mindanao. But there were others to whom Fertig held out the subtle threat of no aid.

For example, he wrote this to another leader in Cotabato just after Pendatun had ceded his authority:

> We are interested in the unification of all guerrilla bands in the province of Cotabato and such unification is to your distinct advantage. Without it you cannot pay your troops, your receipts for food will not be honored by the Army of the United States, your men and your commission will not be recognized. In other words, you will simply be classified as a group of bandits. It is believed that you are an intelligent man, and consequently will make the proper decision.[12]

As will be seen, while the SWPA submarine supply deliveries increased, and Fertig's units received much needed ammunition and weapons, this argument became all the more persuasive, especially as it became increasingly apparent as the months passed that the Americans were coming back.

The work Parsons achieved in consolidating the resistance forces was successful beyond what could have been expected of one man and so few supplies. The important contribution of the submarine visits and the

---

[12] Ltr Colonel Fertig to Major Froilan Matas of October 28, 1943 in 10 MD Correspondence File.

penetration parties was the clear signal it sent to the Filipino people and the guerrillas that "The Aid" would one day come to their country. As such, the belief in the future of the Philippines remained alive, and the Japanese were caused to tie up much needed combat forces to suppress the resistance movement and to stem the trickle of supplies which were reaching the guerrillas.

During Parson's first visit to Fertig, he made it clear that MacArthur was serious about this "lie low" policy. Fertig's units were to make no more attacks on Japanese strongpoints and were to hold only that territory already under guerrilla control. In addition, the guerrillas were not to try to free the prisoners interned in Davao Penal Colony. They were already considered "expended" by GHQ, and failure to free all of the prisoners would likely result in the execution of the remaining prisoners. In any event, Fertig had no idea what he would do with several hundred ill and weakened prisoners if he was able to free them.

Fertig did not fully agree with the orders to avoid combat with the Japanese. The Filipinos had suffered at the hands of the Japanese, and the quiet gathering of intelligence did not satisfy the Filipinos, of which there were many, who now had a blood feud with the Japanese. Moreover, if his guerrillas didn't demonstrate an offensive posture against the Japanese, the people would no longer support his effort, which they had been willing to do at considerable danger to themselves.

Chapter Nine
The Empire Strikes Back

"Guerrilla war, too, inverts one of the main principles of orthodox war, the principle of concentration. Dispersion is an essential condition of survival and success on the guerrilla side, which must never present a target and thus can operate only in minute particles, though these may momentarily coagulate like globules of quicksilver to overwhelm some weakly guarded objective." — B. H. Liddell Hart

Between May 10, 1942 and June 26, 1943, the guerrillas enjoyed a honeymoon away from the Japanese. When American submarines were reported to be in the waters around Mindanao, and reports that they were unloading supplies reached the Japanese, it was only a matter of time before this affront to their Co-Prospeity Sphere would be answered.

So far, the Japanese attacks hadn't penetrated into the interior of the island, rather they contented themselves to looting and then destroying the fields, to starve the people.[1] There were rumors that Morimoto intended a concentrated attack. Fertig's underground indicated that the Japanese attack would come on June 19th. Unless a US offensive begins soon, Fertig wrote in his diary, "we are lost."[2] He knew that he didn't have the forces to successfully engage any kind of serious attack, but he believed in order to save face it was necessary to put up some measure of resistance and evacuate only when under extreme pressure.

The Japanese had already launched a series of attacks along the coast in Cotabato, Agusan, Surigao, Misamis

---

[1] *They Fought Alone*, 221.
[2] Ibid., 224.

Oriental and Lanao. The invading forces had brought along Filipino labor gangs to harvest the crops of rice, corn, coconuts and bananas and to confiscate the fishermen's catches to provide food for the army and to deny food sources to the guerrillas. Reports came into Fertig by runner. The Japanese force was rumored to be 20,000 strong with an additional 150,000 soldiers on the way; gross exaggerations of course. The Japanese command on Mindanao was also at a heightened sense. Ten prisoners escaped from Davao Penal Colony and were assisted by the Mindanao guerrillas after their escape.

Then on June 26, 1943, some 4,000 Japanese struck Misamis City supported by a destroyer and five airplanes. The guerrillas broke and ran, having put up almost no resistance whatever. The capital of the Free Philippines had fallen without a fight. All along the coast towns were deserted, the electricity was out, schools and churches locked, and the banca fleets gone from the bays. Everyone took to the hills. As Father Haggerty wrote:

> But they were wearing us down, pushing us back from fertile land, closing up our lines of communication, blocking us off from the coast, frightening our bancas off the seas. They are killing our work animals, burning all homes, destroying crops. We cannot last if this continues.[3]

However, the Japanese landing was not without its quirk of fate. Among the Japanese who landed at Misamis was a Japanese captain who had once been a high school classmate of Fertig's in La Junta, Colorado. The

---

[3] *Guerrilla Padre*, 129.

Japanese officer talked to Father Healy in the town of Tangub and relayed to Healy that his mission was to find Fertig and kill him. Then oddly enough, the Japanese captain told Healy of how he had once picked cantaloupes with Fertig. He told Healy that he really had no desire to successfully complete his mission and asked him to pass word to Fertig so that he would be aware of the price put upon his head by the Japanese.[4]

All Fertig could do was join his troops in their flight to the hills. His thought at the time was his house-of-cards was coming down around him. Meanwhile, Parsons made a drama-packed forced march across Mindanao to meet the submarine *Thresher* in Pagadian Bay. From there he made his escape to Australia to report on the Philippine resistance movement.

Morgan's Revolt

It has been said, when sorrows come, they come not in single spies, but in battalions. In the wake of the disaster, Fertig observed in his diary: "It is a tragic, comic-opera war."[5] Fertig now had to relocate his headquarters. But to where was the question. He couldn't go to Cotabato, for Salipada Pendatun had still not submitted to his authority, and Fertig would be placed in a very weak position politically by seeking Pendatun's protection. In addition, the Moros were unpredictable and some were known to be working for the Japanese. Zamboanga, though safe, was too far removed for command and control purposes. Davao was controlled by the Japanese, as was Bukidnon.

---

[4] *They Fought Alone*, 240-241.
[5] Ibid., 271.

Agusan had little food, bandit groups still roamed the eastern coast, and the Agusan River Valley offered few advantages to a guerrilla force. But Agusan was under an American commander, Ernest McClish, whom Fertig had yet to meet, and the area offered some advantages for rendezvous' with submarines. In the spring of 1943, an expedition under McClish had cleared Agusan Province of Japanese, although at a heavy price in men and ammunition. They had, however, captured some launches and diesel fuel. At this time of decision, Fertig now estimated that he had some 8,000 guerrillas under his command.

Fertig had realized that he needed an alternate command post should he be killed or his communications with GHQ be destroyed. Then in the fall of 1943, Fertig published the organization's mission to his guerrillas. Their first priority was to gather intelligence for GHQ. Their second priority was to defeat the Japanese on Mindanao. He knew this mission did not fully comply with MacArthur's directive, but Fertig reasoned that without continued public support, the sources of intelligence would dry up.[6]

The Japanese divined MacArthur's "lie low" policy and believed that their punitive expeditions had made the guerrillas less active. They saw the guerrillas as planning for a general uprising on the return of US forces, conserving their forces and building strength, directing efforts towards reconnaissance, and destruction of the lines of communication. In addition, the Japanese had captured intelligence reports prepared by guerrillas for GHQ, and thus had a good idea as to the information GHQ was interested in. For his part, as Fertig observed, "Instructions were to undertake no offensive action

---

[6] Ibid., 199, 307.

against the enemy. This has been followed, but the enemy did not receive the same orders."[7]

Moreover, Fertig couldn't go to Misamis Occidental because he had an inhouse problem there with Morgan. Upon his return to Misamis in June 1943, Morgan expressed his strong dissatisfaction with the Moro truce, the new command arrangement, along with Fertig's orders to avoid contact with the Japanese. In response, Morgan resigned as chief of staff, pulled the men loyal to him out of Fertig's organization, assumed the rank of brigadier general and formed a new command called the "Mindanao and Dutch Indies Command." These were the days of high adventure on Mindanao.

To make matters worse, oddly enough, Morgan's machinations involved even leveraging the Moros to his side. A few weeks after Fertig was blown out of Misamis City, the Datu Busran Kalaw, lord of the Maranao Moros, brought him a piece of paper. On it was a proclamation dated July 4, 1943. Issued while Fertig and his men were on the run, hiding from Morimoto's patrols. The proclamation stated that Morgan had assumed command of the Mindanao and Sulu Islands, replacing Fertig, who had stepped down to become administrative officer of the 10th Military District.[8]

Fertig correctly assessed that Busran's allegiance to Morgan would persist as long as it was advantageous. The next day, a courier arrived with a letter signed by "General Morgan." It told Fertig to appear at a conference to be held in Kolumbugan on August 10.

Sometimes fortune aided Fertig at the critical moment. Fortuitously, a recently arrived submarine had delivered among its military supplies a copy of the May

---

[7] Ltr Fertig to Casey.
[8] Ibid., 273.

31, 1943 *Life Magazine*. The magazine contained a lengthy article on King Ibn Saud, King of Saudi Arabia. The article was replete with pictures and expressions of friendship by the king for the United States. The impact of all this was not lost on the Moro leaders. The Moros were strongly religious in their own way, and the support for America demonstrated by King Saud to the world quickly sealed the decision for the Moro leaders. One hundred or more Moros on Mindanao had actually made the pilgrimage to Mecca, and the meaning behind the interview with King Saud meant infinitely more than any promise Fertig could make. Fertig radioed GHQ to send as many of the magazines to him as they could.[9] In the end, unable to gather an effective force under his control, and unwilling to submit to Fertig's command, Morgan reluctantly board the submarine *Bowfin* bound for Australia. Hedges had wanted Morgan shot.

It was now November 1943, and Fertig moved to Esperanza at the Junction of the Agusan and Wawa Rivers. He remained here until January 1944 when the bamboo telegraph alerted him that the Japanese had become wise to his new location. He pushed farther up-river to Namot Talacogon in the interior Agusan Valley. He remained here until the problem with Morgan was resolved by sending him to Australia and the issue between Pendatun and him was somewhat alleviated.

From then on, conditions began to change rapidly on Mindanao. Coastal towns such as Cagayan in Oriental Misamis were subjected to frequent Japanese patrols and air raids, and the mayor of Cagayan was captured. The fertile lowlands were abandoned and the small barrios along the coast were deserted. Influenza and

---

9 *MacArthur: His Rendezvous with History*, 134.

malaria were killing many, and starvation was becoming acute, especially among the malnourished children. Medical supplies were totally lacking, and amputations had to be done without drugs or anesthetic. One guerrilla, Australian Jock McLaren, performed an appendectomy on himself with only a mirror and a razor blade. The self-conducted operation took five hours to perform. Five days later McLaren took to the jungle just minutes ahead of a Japanese patrol, his appendix in a bottle.

Malaria was the big killer, with deaths on Mindanao estimated at anywhere from 100 to 500 per day. Half of every guerrilla company was down from malaria at any time. Most types of malaria on Mindanao could be treated with quinine or atabrine, and Fertig believed that if every submarine was filled with this medicine it still would not be enough. Benign tertiary malaria was the less virulent strain and was the most common. But by October 1944 there was a full-scale malaria epidemic. Carried by the anopheles mosquito with its notably distinguishable bite (it stood on its head), it was killing its victims within 48 hours.

Lacking sulphuric acid (which could only be imported) needed to jar the quinine crystals from the bark of the cinchona tree, the guerrillas boiled the bark, ground the residue into a powder, and mixed it with coni flour into a pasty pill. The result was a primitive and not very effective medicine. The one positive aspect was that the guerrillas controlled the Del Monte plantation region in Bukidnon Province where Colonel Arthur Fisher, the former director of Forestry, had planted a forest of 11 million cinchona trees before the war. The Japanese were suffering from the epidemic as well, for when Davao City was later captured by American forces, an

Army nurse concluded that twenty percent of the Japanese had malaria.[10]

For the guerrillas, things were tough all over. Two-thirds of the guerrillas were with their families searching for food or they were incapacitated from illness. A sack of corn cost 150 pesos and a sack of rice 250 pesos. A guerrilla earned only 10 pesos a month, and they had not been paid for months because Sam Wilson with his printing presses had been cut off by the Japanese. As it became clearer to the Japanese that an American invasion was imminent, they began to move reinforcements to Mindanao. Japanese patrols no longer numbered only 50 to 100 soldiers.

Now the combat formations were much larger, and they began to move away from the coastal areas and onto the back trails to establish inland garrisons. These soldiers were not the well-disciplined troops that had carried the fight in 1942. Unleashing a new reign of terror, these soldiers "raped, tortured, bayoneted, burned houses and crops, drove off animals, carried away clothing and even plows."[11] The intent of these patrols by the Japanese command was to seal off Fertig and to gain control of the known submarine rendezvous points. And they were meeting their intent.

Far into the interior of the Agusan Valley, Namot Talacogon had become a relatively secure if environmentally inhospitable location for Fertig and his headquarters. On a positive note, the civil government continued to function out of Agusan. The governor and mayors met there, and Father Joseph Luras, Superior of

---

[10] For general discussion see Marco J. Caraccia, "Guerrilla Logistics," U.S. Army War College Student Thesis, 1966, p. 43. For comparison, the malaria rate for U.S. soldiers in SWPA ran as high as 361 per 1000.
[11] Guerrilla Padre, 212.

the Jesuits in Mindanao, conducted his business from there also. The food administration, relief projects, banca coordination and price control board continued to work out of Misamis Occidental. Food was not plentiful, but the headquarters location was secure enough that even a library was established from which reading material, mostly magazines from the submarine deliveries, could be checked out. With administrative efficiency, the "library rules" required that the materials be returned to code "CPZ" via runner within three days of receipt. In this way the guerrillas managed to keep abreast of the progress of the war in Europe and in the Pacific.[12]

As the news of American victories spread throughout the island, the Constabulary soldiers began to defect and the ranks of the guerrillas grew increasingly larger. The Japanese increased their effort and pressed their attack on Fertig's headquarters, an interesting course of action since the Japanese HQ in Manila had already declared him captured and dead. One of the air raids on the guerrilla headquarters was very successful, destroying every radio but one 3BZ which had been buried; communications were nearly wiped out.

It appeared that between 15,000 and 18,000 Japanese soldiers were actively pressing the attacks throughout the Agusan Valley area, and McClish and his 110th Division were under heavy pressure. The attack on the 110th was being spearheaded by the Japanese 41st Infantry Regiment of the 30th Division. The unrelenting pressure drove Fertig farther still up-river to Waloe. But Waloe didn't represent a significant improvement, for the Japanese continued their advance up the river and conducted daily aerial attacks. By Fertig's estimate, his

---

[12] *Guerrilla Padre*, 118.

small force of 300 guerrillas were delaying a Japanese force of approximately 11,000.[13] Fertig and the 110th Division of guerrillas were now fighting for their lives in a jungle swamp much like the Florida Everglades. The local populace for the first time was not friendly to the guerrillas, nor was the surrounding environment. And to top it all off food was running out.

---

[13] *They Fought Alone*, 362-363, 387.

Chapter Ten
Kempeitai, Kalibapi and the Underground

"A general uprising, as we see it, should be nebulous and elusive; its resistance should never materialize as a concrete body." – Clausewitz

As one student of Philippine history wrote of the Japanese occupation policies: "There seems to have been such an unshakable prepossession with Japan's inherent superiority, divine mission, and glorious destiny that the very idea of resistance was intolerable heresy."[1] The magnitude of this intolerance is found in the estimate that one out of every twenty Filipinos died at the hands of the Japanese during the occupation.[2] In today's parlance, the Japanese were either simply not interested in "winning the hearts and minds" of the Filipinos, or they had no clue as to how to go about doing so.

As will be seen, the Japanese approach to counterinsurgency (COIN) on the Island of Mindanao was rudimentary at best. The ratio of Japanese troops to the residents of Mindanao was approximately 1:100. Physically unable to control terrain beyond four major cities with so few troops, the Japanese resorted to terror tactics to subjugate the populace.[3] If any concerted effort was made, the method used could be best described as the "crush them" approach. According to

---

[1] Smith, *Philippine Freedom*, 108.
[2] Day, Shattered Showcase, 106.
[3] Michael Anthony Balis, "The American Influence on the Mindanao Resistance during the Second World War" (master's thesis, Old Dominion University, 1990), 14-15.

James Clancy and Chuck Crosset, if this approach is applied at the outset of a nascent insurgency, it can be annihilated through a vigorous application of force and repression.[4] However, as has been said, the Japanese did not initially consider the guerrillas a military threat. As such, they held back from committing their troops though they maintained some 150,000 men on Mindanao. Moreover, rather than a vigorous application of force, their tactics boiled down to a drawn-out saturation of a particular area with combat troops in the hopes of isolating and eliminating the guerrilla leadership.[5] These tactics, however, rarely worked.

Kempeitai

So, as it always is in insurgency-counterinsurgency warfare, the people themselves are the battlefield. And the most fearsome instrument through which the Japanese dealt with the Filipinos who resisted them was the Kempeitai. The Kempeitai was similar in purpose to the Nazi Gestapo and the Russian KGB, except the Gestapo, vicious as it was, was no match for the efficient terrorism practiced by the Japanese.

Administered by the War Ministry, the Kempeitai was the Army's military police which had full authority for the arrest and "investigation" of civilians and combatants alike. Experts in the application of torture, the Kempeitai, worked with the occupation Army and civil administrators. Together, their tactic was to turn the people against the guerrillas through brute force.[6]

---

[4] As quoted in the Rand Study, *Victory has a Thousand Fathers*, 51.
[5] David Joel Steinberg, Philippine Collaboration in World War II, 1967, 94.
[6] Edward F.L. Russell, *The Knights of Bushido: The Shocking History of Japanese War Atrocities*, 1958, 274-275.

The Kempeitai even had a training manual entitled *Notes for the Interrogation of Prisoners of War*. Two cannon methods for gaining information were "zonification" and "the magic eye." Under "zonification" the Japanese would seize a barrio or town and zone it off (encircle it). The men of the barrio would then be herded into the town square. After proclamations, denunciations, and a general haranguing was finished, a hooded man with eyelets cut in the hood would be brought out – "the magic eye."

This individual, a traitor, spy or purported captured guerrilla, would scan the crowd and pick out spies, guerrillas or sympathizers. These unfortunates selected by the "magic eye" would then be publicly tortured, roasted alive, drawn and quartered, buried alive, or, the end result in most cases, beheaded. Needless to say, the approach of the Japanese to any barrio normally resulted in immediate evacuation and flight to the hills.

On return to their barrios, people would find their homes ransacked or burned to the ground, along with half-eaten animals and fowls left rotting in the street. Adding to these atrocities, it was also a common occurrence for the dead to be dug up in a search for jewelry, and family belongings buried around the houses dug up.[7]

Japanese treatment of prisoners held in their internment camps was also widely known by Filipinos outside the prison walls.[8] The internment camp on Mindanao was the notorious Davao Penal Colony. It remained agonizingly just beyond the reach of the

---

[7] *Guerrilla Padre*, 185-191.

[8] It is recorded that 50,297 Filipino, along with American, British, Australians, Dutch, and Canadians prisoners of war died in the internment camps.

Mindanao guerrillas. In the period April 1942 to September 1942, no fewer than 200 prisoners were buried each day in this Japanese internment camp; 27,000 were buried in all.[9] In these camps, Filipino prisoners generally fared even worse than American or Allied prisoners, presumably because their "heresy" to the Japanese cause.[10] The Japanese were in fact particularly piqued by the protection provided by the Filipinos to the Americans still remaining unsurrendered throughout the islands.

An indication of the severity and extent of Japanese methods in dealing with the guerrillas can be observed in the way they dealt with captured guerrillas. According to their ancient code of Bushido, captured Americans and Filipinos were "captives of war," not prisoners of war. As such, they were told that they could expect no clemency. And as merely bandits, they could be killed outright.[11] Captured guerrillas were tried and summarily executed on charges of "baneful action."

In all of this, the Japanese certainly ignored the wisdom of Sun Tzu, whose writings had served the Samurai well. Sun Tzu had written, "Treat captives well, and care for them...All soldiers taken must be cared for with magnanimity and sincerity so that they may be used by us."[12]

Their repressive measures having failed, in December 1943, the Japanese turned to an amnesty approach and published the following proclamation:

---

[9] *Guerrilla Padre*, 116.
[10] Robert Ross Smith, "The Status of Members of Philippine Military Forces During World War II," 1973, p. 20.
[11] *Guerrilla Padre*, 117.
[12] Sun Tzu, *The Art of War*, translated by Samuel B. Griffith, 1963, 76.

> The amnesty under which Americans have been guaranteed safety and internment by the Imperial Japanese Government is about to expire. After January 25, 1944, any American found in the Islands, whether unsurrendered soldier or civilian, will be summarily executed.[13]

The Japanese were true to their word, for aggressive Japanese patrols hunted down American families hiding in the mountains and killed many prior to the January 25th deadline. This proclamation led General MacArthur to drastically increase his efforts to extract by submarine Americans who were not fighting with the guerrillas. This effort, coordinated by Fertig on Mindanao for the most part, is generally viewed as one of the most significant contributions the guerrillas made during the war.

In view of the Japanese COIN approach as a whole, the bankruptcy of their policies merely served to bolster the guerrilla ranks and fanned into flame the intense hatred of the Filipino people for their Japanese captors.

In August 1943, the Japanese tried a new approach. To win the people, they dangled the carrot of an independent Filipino government and independence. Under this approach, the Japanese hoped that jurisdiction over the guerrillas would revert to the puppet government. In this way, the Filipino guerrillas would no longer be combatants but traitors to their own government. They would just be common bandits.[14]

In October, 1943, in a further move to placate the guerrillas, the newly installed puppet president Jose P. Laurel declared amnesty for all political prisoners and set free thousands of imprisoned guerrillas. The government further declared amnesty for all guerrillas

---

[13] *Guerrilla Padre*, 160.
[14] Gene Hanrahan, Japanese Operations Against Guerrilla Forces, 1954, 17.

and set a January 20, 1944 deadline for their surrender, after which, as Laurel put it, "the government would take drastic action to force the guerrillas to surrender."[15]

The sham fooled no one for the Japanese still ran the country and none of the freedoms associated with independence, such as free elections, free press, and so on were permitted. Moreover, the puppet government demonstrated no resolution to carry out the threat, and the guerrillas, sensing the weakness of the Laurel government, simply ignored it. Thus, the brutal treatment of the population by the Japanese occupiers continued as usual.

The consequence of these actions, and one that was secretly supported by some members of the Laurel government, was that the newly-freed prisoners, having survived the brutality of the internment camps, joined the guerrillas. In fact, during the 120 day "pacification program," many Filipino leaders publicly protested against the guerrillas but privately aided the resistance movement. The three-month hiatus gave the guerrillas an opportunity to regroup and train their new members. However, the program was given the appearance of success because lists of fictitious names and unserviceable firearms were "surrendered" to the local Filipino authorities. At least on paper the amnesty program was working.

On Mindanao, the stand down in activity came at a critical time. Fertig had reached a critical point in his organizing of the Mindanao guerrillas, having just weathered a three-month period during which the Japanese had mounted a concerted offensive against his Misamis City headquarters.

---

[15] Office of Strategic Services, "The Program of Japan in the Philippines," pp. 272-273.

Kalibapi

While the Japanese didn't initially consider the guerrillas a military threat, they did, however, fear the power gained by the guerrillas by way of the tacit approval of the population. With their COIN approaches failing in succession, the Japanese sought a new way to deal with the guerrillas, this time through a political organization – the Kalibapi. Launched in December 1942, while advertised as the exclusive political party for the Philippines, the Kalibapi in reality was a propaganda arm for the Japanese Military Administration. It therefore served to gather intelligence on local populations to assist in their monitoring and control.

As the political arm of the puppet government, the Kalibapi was tied into the Japanese-established "Neighborhood Associations." The "Neighborhood Associations" were given the responsibility for distributing food and commodities. In this way, the Japanese not only determined who would receive food, they also created a spy network for rooting out guerrilla sympathizers.

In this manner the Japanese sought to control the movement of guerrillas in and out of the barrios and hamper recruiting efforts. The long arm of the Kalibapi and the Neighborhood Association placed a great strain upon the guerrillas and civilian population alike. Both could be extremely hard on collaborators, and the issue was a very, complex and personal one with which the guerrilla leaders had to contend. It is thought that between seventy and ninety percent of the population as a whole resisted the Japanese in whatever way possible.

And while most resisted, Filipinos could be known to collaborate with the Japanese for any number of reasons, principally it was out of the desire to protect one's family and personal interests. Most collaborators were the elites in Manila, or at least they received the greatest attention for having collaborated.[16]

On Mindanao, the collaboration issue was most often settled in the trenches, and the guerrillas had a good idea of who had supported them and who had not. Fertig himself was no hardliner on collaboration. He believed that Filipinos who subjected themselves to the Japanese treatment in order to protect their own people were courageous in their own right. He distinguished these people from the traitors who purposely aided the Japanese in their fight against the guerrillas. And, of course, Mindanao was populated in the interior by tribes who had no idea who the Japanese and Americans even were, and any assistance they may have rendered to either side was apolitical from the standpoint of the resistance movement itself.[17]

Important to the resistance movement throughout the Philippines, and critical to the guerrilla effort on Mindanao, was the Filipinos' attitude towards the Americans. The persistent irony was that the Filipino guerrillas had established contact with, urged leadership upon, and collaborated with a people who had been the past conquerors of their country. Unlike most other Asian peoples who had been sympathetic to Japan's

---

[16] Bernstein, *Philippine Story*, 170-171. See also Ira Wolfert, American Guerrilla in the Philippines, 1945, p. 84.

[17] The impending American invasion of the Philippines became something more than a matter of ridding the islands of the Japanese, for the nation now had to deal with its collaborators. General MacArthur initially "posited general culpability until the individual could establish his innocence; the Osmeña policy Osmeña succeeded Quezon on his death posited individual innocence until treasonable motivation could be assessed." D. Clayton James, The Years of MacArthur, 1941-1945, Volume II, 1975, p. 529.

"Asia for the Asians" theme, the Filipinos did not react the way the Japanese had expected them to react.

In November 1943, President Laurel, who called the guerrillas fools and renegades, decided to increase the size of the Philippine Constabulary to 40,000. It had previously been reconstituted given limited authority back in May 1942. The hope was that the Constabulary would bring legitimacy to the Japanese counterguerrilla program and that they would eventually be used against the Americans as a conventional force.

The irony here is that the Japanese tried to use the Philippine Constabulary for the same purpose for which it was originally formed by the Americans: to pursue and eliminate guerrilla bands. All parties were happy with this decision. For the Japanese, the Constabulary would free Japanese troops for other duty. As for MacArthur, he believed that the brutal treatment of the Filipino people would lessen, and the guerrillas, they saw in the Constabulary an impotent if not neutral foe.

For this purpose, Laurel recruited or impressed many former officers and noncommissioned officers of the Constabulary back into service. Major General Paulino Santos, a former Constabulary officer and Governor of Lanao was appointed as Commissioner of Mindanao with quasi-military authority over the island.

In the last analysis, the Philippine Constabulary had limited success in controlling guerrilla activity. In response, on June 10, 1944 Laurel created the Bureau of Investigation within the Constabulary, apparently to mimic Japanese practices. This bureau had the powers and mission of the Kempeitai. However, by July, 1944, knowing that the experiment had failed, Laurel had published Proclamation Number 20 which, in effect, reverted responsibility for dealing with the guerrillas back to the Japanese.

By October 1944, the Japanese began to disarm, disband and intern the Constabulary members. However, the guerrillas made good use of the Constabulary while it existed. It was a ready source of weapons, whether turned over conspiratorially or won through military action.

The Underground

The Filipino early response to the Japanese had been sufficiently vexing to cause General Masaharu Homma, Commanding General of the Fourteenth Army, to declare that Filipinos with "Pro-American sympathies" would "be annihilated without mercy."[18] As the Japanese policies became increasingly counterproductive, the Japanese propaganda became more penetratingly shrill. According to the Japanese, Filipinos were no longer "true Orientals" but had become a hybrid of American political principles coupled with inherited Spanish values and Oriental philosophy.[19]

While the guerrillas survived, the Mindanao underground turned out a flow of information. This kept the resistance alive by enabling it to stay at least one step ahead of the Japanese occupation forces, especially the Kempeitai. The 10th Military District's web of underground agents canvased the island, burying themselves into even the most unlikely of positions. For instance, one of Hedges' agents was a young lady who worked in the Kempeitai office in Iligan. She kept the

---

[18] *The Fought Alone*, 208-209.
[19] Steinberg, Philippine Collaboration, 57.

105th Division well supplied with all the latest, warning the guerrillas of intended patrols.[20]

Perhaps the most unique source of information for Fertig was the mistresses of the ranking Japanese officers. Before General Homma left the Philippines, his mistress was a good source of information. And one of Fertig's couriers had a cousin who was the Filipino lover of General Morimoto's mistress.[21]

Ultimately, the Japanese never tried to understand the Filipinos, or at the very least be inclined to tailor their occupation policy to the Filipino culture. The Americans had respected the Filipino dignity, and that had become the true basis for American power in the Philippines. It would now pay large dividends. The Filipino people would help Americans who were complete strangers, providing the best food and shelter available to them. They would transport gravely ill Americans for miles over treacherous mountain trails to keep them from falling into the hands of the Japanese. Often, the visit of an American to a village was enough reason in itself for the village to have a fiesta. America in the Philippines was supported by the common citizen, a fact with which the Japanese were unable to come to grips.

---

[20] Robert B. Osprey, *War in the Shadows: The Guerrilla in History*, Vol 1 (New York: Double Day, 1975), 519.
[21] Keats, They Fought Alone, pp. 135, 203.

Chapter Eleven
Beans, Bullets and Submarines

"Guerrilla war is waged by the few but dependent on the support of the many." – B. H. Liddell Hart

The practical forms of association between a regular force and the resistance movement it supports are several. The most important form of support is logistical supply, especially supply of weapons, ammunition and medicine. Food may be sent to the guerrillas though it is usually the most difficult to supply. The sending of officers and agents to act as liaison personnel, training advisers, or perhaps to be actual commanders are other forms. The introduction of non-indigenous personnel into the area of operations, however, can create more problems than it would solve.

The support given to the Philippine resistance movement by GHQ, SWPA has been viewed as crucial to the success of the movement. In return for the aid, SWPA received an abundance of intelligence which facilitated naval operations in the Southwest Pacific and assisted in the planning for the invasion of the Philippines. Irrespective of who benefitted the most from the support, it is certain that the assistance was important for it provided some measure of usable military materiel and served as a symbol of hope to the Filipinos in a most dramatic way.

In August 1943, Fertig wrote a letter to Datu Gumbay Piang which characterizes well the logistic problems faced by the Mindanao guerrillas: "You will note that it is necessary for every applicant for induction in the Maguindanao Militia Force to be armed. We do not furnish the arms; the men furnish their own."[1] He had written to GHQ two months earlier that he desperately needed supplies to fight the Japanese and to demonstrate the US intent on coming back to the Philippines else the civilians will succumb to the cumulative effects of Japanese propaganda. Fertig didn't really care how the supplies arrived on Mindanao, although he did believe as early as July 1943 that aerial resupply was feasible. He had gotten this idea from William E. Dyess, an aviator who had served on Bataan and later escaped from Davao Penal Colony.

Fertig himself had overseen the construction of airfields on Mindanao so he knew what the guerrilla capabilities might be. He wrote to General Hugh Casey that he didn't believe there were technical difficulties which could not be overcome if the desire were there. As it was, aerial supply didn't begin in the Philippines until after the Leyte landings in October 1944. But once this means of resupply was opened, Mindanao, Cebu and Panay would receive approximately 280 of the 500 sorties flown to the Philippines.

Spyron

All things being equal, however, the method used to bring supplies to the guerrillas was the submarine. While on Corregidor, MacArthur had maintained a top-secret file code-named "George." The folder contained his plans for conducting guerrilla warfare in the Philippines.[2] Although the guerrilla movement never got

---

[1] Ltr Col W. W. Fertig, HQ 10 MD Office of CG In the Field to Datu Gumbay Piang, Cotabato Area of Aug 4, 1943, in Correspondence Files, 10th Military District.
[2] Edward Dissette and H. C. Adamson, Guerrilla Submarines, 1972, 13-16.

started as he had envisioned, ideas were beginning to form in his mind on how a guerrilla movement might be sustained. At that time submarines were running the Japanese pickets and minefields to resupply the beleaguered Bataan-Corregidor garrison. It required only a short step to envision submarines resupplying the guerrillas.

Thus "Spyron" was born, one of the closest guarded secrets of World War II. Essentially a one-man operation at its inception, "Spy Squadron" was a code name for Commander Parsons who initially established the Special Mission Unit that provided the supplies and submarines for the clandestine supply drops in the Philippines. The Spyron activities lasted two years, during which time supplies were delivered, evacuees rescued, and special intelligence teams were landed in the Philippines.

US Navy records indicate that of the submarine resupply missions to the Philippines officially conducted, sixteen visits were made to Mindanao by six different submarines: *Narwhal* - 9; *Trout* - 2; *Stingray* - 2; *Nautilus* - 1; *Tambor* - 1; and *Bowfin* -1. The *Narwhal* sailed the first mission and the *Stingray* made the last on January 1, 1945.[3] Only one submarine was lost during these two years, the *Seawolf*, which went down enroute to Samar in September 1944. The Special Mission Unit was dissolved after the last run in January 1945.

The metrics are impressive. During the two-year period, 1325 tons of supplies and equipment were delivered with none falling into enemy hands before or

---

3 U.S. Seventh Fleet, Memorandum, subject: "Submarine Activities Connected with Guerrilla Organizations in the Philippines," no date (hereinafter 7th FLT Memo).

at delivery. Three hundred thirty-one persons were landed and 472 evacuated. All told, 19 different submarines in all carried out Spyron missions in the Philippines.

MacArthur wanted absolute secrecy for his submarine deliveries. The training camps in Australia where Filipino recruits from the 1st Filipino Infantry Regiment and the 2d Filipino Infantry Battalion in California were indoctrinated and trained for intelligence teams was closely guarded. The submarine operations were kept so secret that even the penetration parties did not know the date of their departure. An agent would be roused from bed, put into a pair of dungarees, and joined with a labor crew loading a submarine in the middle of the night.

Then on one trip into the submarine, he would be kept inside and another agent would leave the submarine to take his place in the labor crew. His personal gear would be brought aboard the vessel for him later.[4] In this manner the coastwatcher teams and special intelligence agents made their way to the Philippines.

The first submarines used in the resupply missions were patrol submarines (attack submarines in current terminology). As these submarines left Australia for missions in the Western Pacific they would stop briefly off the Philippine coast and discharge a small amount of cargo or a small penetration party by rubber boats. The 7th Fleet was not enthusiastic about using its few submarines to support what appeared to be at best marginally important missions, particularly when the missions called for surfacing in enemy waters virtually under enemy guns. In addition, any cargo or personnel transported meant the removal of one torpedo for every

---

[4] See Ind, AIB, p. 123; Dissette, Guerrilla Submarines, 65; Mellnik, Philippine Diary, p. 285; OCMH, "Resistance Movement," 227-228.

man and his gear that came aboard the submarine (the maximum load was six of these Spyron passengers).

The benefits of having coastwatchers in the Philippines soon became evident, however. Reports from the coastwatchers on weather conditions and on enemy ship and air movements soon caught the Navy's attention. One routine report in particular from an observer near Davao City caught their attention. A Navy submarine had sunk a Japanese ship, and this had been routinely reported by the observer.

The Navy was interested in the report because with their hit and run tactics submarines often did not have the opportunity to confirm their kills and to assess the damage on their targets. The coastwatchers could do this from their mountaintop hideouts. The equation was straightforward. To improve survivability, the Navy needed coastwatchers. The coastwatchers and radio teams needed guerrilla protection. The guerrillas needed weapons, ammunition, radios and medicine – which meant, of course, that they needed submarines.

Initially the patrol submarines *Seawolf* and *Stingray* were assigned to GHQ, and in September 1943 Navy assigned the *Narwhal* to support the Philippine Regional Section of GHQ. The *Nautilus*, the *Narwhal's* sister boat, was reassigned from Pearl Harbor to Brisbane on May 3, 1944 and assigned to support GHQ.[5] The importance of these two vessels is that they were both cargo-carrying submarines. The small fleet-type submarines could carry from five to ten tons of supplies, and their use was dependent upon the submarine's operational schedule. The cargo-carrying submarine: with their larger 3,000-ton displacement had dedicated

---

[5] General Headquarters, Far East Command, Operations of the Allied Intelligence Bureau, 1948.

supply missions and could transport from 50 to 100 tons of supplies.

The *Narwhal* and *Nautilus* were commissioned in 1930 and were called "super subs." With speeds of 17 knots on the surface and eight knots submerged, the submarines were large cumbersome workhorses. Their diving time was not rapid, and their torpedo capacity was limited. The boats had ten torpedo tubes, carried 26 torpedoes, had two six-inch deck guns and machine guns. Each had a complement of eight officers and eighty men. The crew and torpedoes carried could be reduced to permit the submarine to comfortably carry ninety-two tons of supplies. Before coming to Australia, the submarines had been used to land Marines and Army scouts, evacuate civilians and conduct photographic reconnaissance elsewhere in the Pacific.

The "super subs" were noted for their engineering problems, and the *Nautilus* had been undergoing overhaul which had delayed her earlier assignment to the Special Mission Unit within the Philippine Regional Bureau. The *Narwhal* was referred to affectionately by her crew as "Inchcliff Castle Maru" after a fictional run-down tramp steamer with noisy engines in a then currently popular *Saturday Evening Post* series. The *Narwhal* under-went overhaul in the spring of 1944.

When she put to sea again on her first trip she "hogged," bent in the middle, presumably from loading too much cargo in her torpedo rooms. By summer 1944 she was even more noisy and was leaving an oil slick and a smoke trail. Her best condition for operating was heavy weather.[6]

---

[6] U.S. Naval Institute, Reminiscences of Rear Admiral Arthur H. McCollum, U.S. Navy. (Retired1, 1973, Tape 12, 560; Dissette, Guerrilla Submarines, 23, 80-82, 95, 130, 132-133.

The first submarine to be sent by GHQ, the *Gudgeon*, was to go to Pagidian Bay, Mindanao to land Major Jesus Villamor and a radio team. Villamor carried microfilm with a high-grade cipher system for both Fertig on Mindanao and Peralta on Panay in a dental alteration and under a patch on his gym shoes. The submarine was unable to rendezvous at Mindanao so it made for Negros and dropped the landing party there.[7] GHQ followed up this attempt with a second, successful one. Charlie Smith and Chick Parsons' "Fifty Party" arrived in Pagadian Bay four miles southeast of Labangan board the *Tambor*.

Beans and Bullets

The submarine deliveries which arrived on Mindanao were diverse but limited by the capacity of the submarines, the diameter of the submarine hatch (23 inches), and the mission of the guerrillas. Items such as radios were broken down into component parts, and military supplies, magazines, clothing, chocolate bars, soap and cigarettes were put into waterproof tins. Popular items were sewing kits, pencils and shortwave receivers to be smuggled into the villages and towns occupied by the Japanese. If caught by the Kempeitai, death was the punishment for possessing a radio receiver.

Although viewed as frivolous by GHQ, the guerrillas pleaded for heavy duty sewing machine needles. Every village had a sewing machine and a generator, and the heavy burlap material used for dresses was wearing out the sewing machine needles. The subs also brought plates for printing money and paper arrived on Mindanao. And to boost morale, oranges sent to the

---

[7] Ind, AIB, pp. 122, 125-126.

island were stamped "USA" whether they were grown there or not. Candy wrappers and matchbook covers were stamped "ISRM" and "I Shall Return MacArthur" as were cigarette packages, chewing gum wrappers, pencils, and other small items. Magazines with news of America's victories were also brought in large number.

These items developed considerable propaganda value, and they would often turn up on the desks of Japanese officers. Five-gallon kerosene tins were three-quarters filled with wheat flour used to make communion wafers, and a half-dozen bottles of Mass wine filled the remaining fourth. Religious medals, candles, rosaries and holy pictures were wedged in among the bottles. These supplies were highly valued because there was no wheat for making flour or grapes for making wine in the Philippines. Priests used eyedroppers to serve the wine at communion.

Among the most important supplies were cathartic pills, sulfathiazole, and atabrine and quinine for combatting malaria. Parsons even took pains to ensure that the guerrilla leaders' pride was assuaged. Pendatun received saddle soap for his shiny cavalry boots, Kangleon on Leyte received adhesive for his dentures (his gums had shrunk from malnutrition resulting from his internment), and Fertig received six bars of soap.[8]

Things did not always go just right, of course. On one occasion, the guerrillas unloaded 20 mm guns but no shells for them. They were told the shells would be in the next submarine shipment. As it turned out, while the Navy had the guns, the Army had the rounds. Hammocks with zippers arrived, but ammunition was far more useful than a hammock in a surprise attack. On

---

[8] Ind, AID, 164; Wise, Secret Mission, 136-137; Edward Haggerty, Guerrilla Padre, 1946, 23.

another occasion, several cases arrived with markings indicating "submachine guns." The cases actually contained cavalry sabers. Moreover, to the guerrillas chagrin, boxes of whiskey and pesos didn't always make it through the long supply line. However, GHQ began marking these boxes "military rations," and the problem ceased.[9]

A shipment of supplies brought in by patrol submarine might have these items:[10]

6 radio sets	4 batteries
10 cases .45 caliber ammo	12 cases .50 ammo
10 cases hand grenades	13 Tommy guns .45 caliber
1 air-cooled .50 caliber MG	3 Springfield rifles
Several cases personal items medical supplies	

By March 1944 the breakdown of supplies by type within a shipment, listed by priority of importance to the guerrillas, would look like this:[11]

Percent	Item
60	Ordnance
8	Medical
10	Signal and engineering
5	Currency requirements (paper, ink, plates)
10	Sundries (specific requisitions)
7	QM items and propaganda items

It goes without saying that the supplies brought in by submarine gave new life and hope to the guerrillas throughout the Philippines. More than that, it may be

---

[9] *They Fought Alone*, 330, 342.
[10] Dormal, The War in Panay: A Documentary History of the Resistance Movement in Panay-u World War II, 1952, p. 87.
[11] Headquarters, Philippines Command, "U.S. Army Recognition Program of Philippine Guerrillas," no date, p. 234, Appendix I, p. 4.

argued that without the submarines the guerrilla movement would have collapsed long before Allied forces could return to the Islands.[12] In addition to supplies, submarines undoubtedly facilitated the early return of the invasion force to the Philippines by bearing along the coastwatchers trained in Australia and the radios that transmitted the intelligence information on the Japanese strength and dispositions.

This was the aim of Spyron all along. More than that, the goal was to put a rifle in the hands of every guerrilla by the time the invasion force reached the Philippines.[13] This goal was achieved, and some stories of the success of the supply missions are apocryphal:

> How extensive this work of the submarine was I learned one day when two ladies of the underground came to see me. I outlined Saturnino's project of raiding Muntinlupa for the 2,000 rifles stored there. They only stared at me; they almost sniffed. "Only 2,000 rifles! Why, that is nothing. It is not worth the risk."[14]

However, the supplies sent to the guerrillas have also been described as a "mere bagatelle," and that characterization is supportable if we look only at the raw figures. Sources put the tonnage of supplies shipped by submarine to the Philippines at between 1,325 tons and 1,600 tons.[15] If we use the figure for the number of guerrillas actually recognized as having fought in the

---

[12] Dissette, *Guerrilla Submarines*, 234.
[13] Ingham, *Rendezvous by Submarine*, 199; Wise, *Secret Mission*, 145
[14] Forbes Monaghan, *Under the Red Sun: A Letter from Manila*, 1946, 223.
[15] The Navy claims 1,325 tons; see 7th FLT Memo. The 1,600-ton figure comes from an Army Department historian, see Department of the Army, Office of the Chief of Military History, "The Philippine Guerrilla Resistance Movement," no date, p. 216 (hereinafter OCMH, "Resistance Movement").

resistance movement, 260,715, and the more liberal figure for tonnage, 1,600, then we have a per guerrilla supply rate of 12.3 pounds per guerrilla for the two year period over which submarines had been making deliveries, or just over six pounds per year. That is not much once the chewing gum, magazines, packaging materials and the like are eliminated. Tonnage sent to Mindanao is reported to have been 750 tons.[16]

Much of this was cached on MacArthur's orders and much of it was sent to the islands to the north, so the Mindanao guerrillas did not use nearly this amount. Fertig's guerrilla strength was estimated at between 25,000 to 40,000, with most sources agreeing on 36,000. If we use the accepted strength figure along with the 750-ton figure for supplies sent to Mindanao, we have a figure of 41.6 pounds per guerrilla between March 1943 and September 1944 when the last submarine put in at Mindanao.

Regardless of how the allocations per guerrilla are computed, the conclusion is the same. The Mindanao guerrillas had by far the largest allocation of supplies per guerrilla, but even this tonnage does not suggest an adequate resupply rate to the guerrillas. If the 41.6 pounds per guerrilla were ammunition alone, which it was not, it was barely enough to get him through a couple of good fire fights. It is obvious that the guerrillas fought mostly with heart and with the hope brought by the submarines.

The distribution of the supplies was a different problem. GHQ, SWPA generally regarded the Mindanao guerrilla movement as the best organized of the guerrilla organizations in the Philippines and as the center of resistance in the archipelago. Furthermore, MacArthur's

---

[16] ASI, "Role of Airpower," p. 140.

plans called for returning to the Philippines with Mindanao as his first objective in the islands and subsequently using the island as his toehold for the push north to Manila. Mindanao had much to recommend it as a location for pre-positioned supply storage, for Mindanao was relatively convenient to MacArthur's supply lines and the island was not heavily garrisoned with Japanese.[17]

MacArthur had written Fertig a letter directing him to act as his supply point for furnishing supplies to the other islands. Fertig was also to hide stores of supplies on Mindanao for future use by the invasion force and for the guerrillas to use when the invasion force arrived. He was not to use the weapons against the Japanese so as not to invite retaliation and increased Japanese anti-submarine activity.[18]

Fiesta

Ultimately, Fertig was directed to keep submarine deliveries absolutely secret. And while Fertig could keep the arrival time and location of the submarine rendezvous secret, once it surfaced, he couldn't keep the arrival of the submarine much of a secret. As it turned out, a submarine's arrival was cause for great rejoicing among the Filipinos on Mindanao. In fact, they celebrated its coming with fiestas. These celebration activities had a salutary effect on the morale of the guerrillas and strengthened the resistance movement as a whole. In any case, with "ISRM" matchbooks and

---

[17] Supreme Commander for the Allied Powers, Reports of General MacArthur: The Campaigns of MacArthur in the Pacific, Volume I, 1966, p. 309; Uldarico Baclagon, Philippine Campaigns, 1952, pp. 302-304; Reminiscences - McCollum, Tape 13, p. 596; Ind, AIB, p. 125.
[18] Keats, They Fought Alone, p. 332.

candy wrappers turning up frequently on the desks in Japanese Army offices, the conclusion to be drawn by the Japanese was pretty evident.

After the fiesta, a lot of work went into distributing the cargo. It called for extensive planning on the part of both Fertig and his guerrilla commanders. Cargadores had to be brought together in sufficient numbers to carry the supplies but not so soon as to betray the pending rendezvous. A banca fleet and lighters had to be gathered, again under the same stipulations. Supply routes through the mountains had to be planned for and then given additional security where needed.

A normal submarine load of supplies would take 3,600 cargadores to pack the supplies across the mountains if each man carried 50 pounds and carabaos were not used. The supply column would have extra cargadores and a security force, so put the column at about 4,000 men. If they had a march of 14 days, seven days each way, and they ate two meals a day, the column would require 112,000 meals which would have to be carried with the column or provided for along the way. And, of course, each cargadore would have to be paid for his labor. With this in mind, logistical problems could mount rapidly.

By the spring of 1944, the flow of supplies to Mindanao was the buzz around the island. The Japanese learned of the submarine deliveries primarily from their agents. Their intelligence reports showed a continuing and growing concern. They had inflated notions of what the submarines were accomplishing, for their reports speak of guerrillas unloading "parts of five airplanes" from one submarine. They also believed that the submarines were apparently bringing in field artillery pieces and anti-aircraft guns.

Figure 11-1. Submarine Rendezvous Points.

They thought that the delivery of small arms ammunition was plentiful and that each guerrilla carried between 20 and 30 rounds of rifle ammunition. They fixed the favorite rendezvous locations used by the guerrillas very quickly and accurately and sent large forces into these areas to control them. The Japanese focused especially on Pagadlan Bay, Butuan Bay and the east coast. As the submarine visits increased, the Japanese tracked an increase in the guerrilla activity. They believed that a large number of Americans were being landed on Mindanao by submarine. By mid-1944, 18 percent of the Japanese submarine sightings were off the Mindanao coast. Figure 11-1 above depicts the actual rendezvous locations at Mindanao between March 1943 and September 1944.

Claude Fertig

Along with this new Japanese attention to the submarine deliveries came an unwelcome change in Japanese policy on the treatment of Americans. It went from bad to much worse. The Japanese proclaimed that any unsurrendered American found in the islands would be executed on the spot after January 25, 1944.[19] They meant it, and a relentless search for Americans hiding in the mountains commenced. Peralta radioed to Fertig before January 25th, and therefore, before the so called "amnesty" period was up: "Report thirteen American nationals, among then women and children, have just been slaughtered by the Japanese on Panay."[20] Fertig's brother Claude and his wife who was eight months pregnant were on Panay, and they had barely escaped by seconds the Japanese patrol which had killed the thirteen Americans and their Filipino friends and workers.

On being notified of this new situation, MacArthur reassigned the patrol submarines to an evacuation role, and Fertig coordinated many of the evacuations by radio from his headquarters. In all, 472 people were evacuated. The condition of many was pitiful. Captain Olsen of the *Angler* picked up 58 evacuees in March 1944 from Panay, among them Claude Fertig, his wife and his newly-born daughter. Olsen had this to say of their condition: "The ship was immediately infested with cockroaches and body and hair lice. A large percentage

---

[19] Ingham, *Rendezvous by Submarines*, 181.
[20] Ibid., 182.

of passengers had tropical ulcers, plus an odor that was unique in its intensity." He called the boat's compartment where the people were billeted the "Black hole of Calcutta."[21]

The submarines themselves were always subject to danger on their guerrilla supply missions. As has been said, the Filipinos would have a fiesta on the arrival of a submarine, and at short notice. Many a submarine commander felt his heart in his mouth when he surfaced to carry out his super-secret mission only to find a bar and dancing girls waiting on the shore.[22] The Japanese were rarely more than a few miles away, so the captain's misapprehension was well-founded. In one instance, Commander Latta took the *Narwhal* right up to a pier at Nasipit in Butuan Bay, probably the only time an American warship had tied up to a pier to discharge cargo in enemy territory.[23] The *Narwhal* was in a later visit to proceed up the Agusan River itself to discharge cargo and pick up thirty-two evacuees. She beached, and with the Japanese only three miles away, the crew commenced "sallying"– running fore and aft the length of the ship – to rock the boat from the sandy shoal upon which she was lodged.[24]

If a submarine was unable to keep a rendezvous, things could become difficult very quickly. The boat could steam to another location if need be, but the cargadore and banca fleets could not be shifted quickly at all. And if dignitaries had been invited to see the

---

[21] Dissette, *Guerrilla Submarines*, 113-116.
[22] Reminiscences - McCollum, Tape 13, p. 601. Hawkins, Never Say Die, pp. 193-194.
[23] *They Fought Alone*, 327.
[24] Dissette, *Guerrilla Submarines*, 88.

submarine rendezvous, then the guerrilla leader could "lose face" quickly as well.[25]

As a general rule, World War II submarines spent about 90 percent of their time on the surface and submerged only to avoid attack. Sometimes they would lie off the coast of Mindanao submerged waiting for the signal from the guerrillas ashore or just observing the rendezvous area. In fact, the water was so clear in places that on a moonlit night, and often during the daylight hours, a person could sit ashore, especially on a hill, and see a huge dark object resting on the white sand, a submarine.[26]

To protect the submarines, the guerrillas would go to great lengths. For example, when Parsons left Mindanao on July 15, 1943 he sat in a boat which "resembled a greenhouse." It was piled high with petted palms lashed to the railings, banana leaves hanging from the guy wires and a coconut tree strapped to the mast. It was supposed to look like a small island in Pagadian Bay.[27]

On occasion the submarines ran afoul of the Japanese and had to defend themselves. The *Narwhal* had several close calls near Mindanao but managed to survive Japanese depth charges.[28] The submarines were also vulnerable to air attack because they spent so much time on the surface. A friendly air patrol in the area almost always improved the security of a submarine. But towards the end of the war, the submarines were not necessarily any more secure with American fliers aloft. The only submarine to be lost supporting the guerrillas was the *Seawolf* which was mistakenly sunk by Navy

---

[25] Ibid., 100; *They Fought Alone*, 321.
[26] *Guerrilla Submarines*, 84.
[27] Mellnik, *Philippine Diary*, 274-275.
[28] For example, see *Guerrilla Submarines*, 94, 101-103, 107; Ind, AIB, pp. 195-196.

pilots from the carrier *Midway* while the *Seawolf* was steaming from Australia to Samar.[29]

In a rare instance, an agent from the War Department, Captain Harold Rosenquist, managed to make it to Mindanao through the good offices of Major Steve Mellnik and General Sutherland. Mellnik had escaped from Davao Penal Colony, and Rosenquist was in the business of springing POW's in Europe. Mellnik had met Rosenquist at a debriefing at the Pentagon, where Rosenquist became intrigued with the possibilities for the liberation of the Davao prisoners.

The Japanese kept 2,000 American prisoners in the penal colony. When twelve men escaped, later joining Fertig's guerrillas, the Japanese beheaded twenty-five prisoners. Mellnik inserted the OSS S-X intelligence officer Captain Harold Rosenquist into Mindanao in an attempt to rescue the Americans before they could be moved. However, the Japanese had already evacuated the camp, placing the American prisoners on a ship bound for Japan. Tragically, that ship was sunk by an American submarine, and only eighty-three reached shore to be rescued by guerrillas.

All told, 472 Americans were evacuated from Mindanao by submarine. One of the most gruesome evacuations took place on Mindanao and stemmed from an event that took place just off Sindangan Point near Sindangan Bay on September 6, 1944. By the Summer of 1944, the Japanese had begun, transporting American prisoners from the Philippines to labor camps in Japan, Korea and Manchuria. The ships were of all types, from military troop ships to merchant vessels. On September 7th, five ships of a seven-ship convoy transporting American prisoners were sunk.

---

[29] *Guerrilla Submarines*, 159-160.

As the vessels sank, the Japanese guards gunned down the prisoners in the holds from the hatches above. Many who did manage to escape from the ships were shot or clubbed to death in the water by hysterical Japanese guards. Only 80 of the approximately 800 prisoners survived from the five ships, and they made their way ashore at Sindangan Point. Five hundred corpses washed ashore with them. There they were picked up by guerrillas, and later those who were still alive were put on the *Narwhal* near Siari Bay.

Chapter Twelve
Cargadores, Coconut Oil and Cannons

"The prospects and progress of a guerrilla movement depend on the attitude of the people and their willingness to aid it by providing information and supplies." – B. H. Liddell Hart

Providing adequate logistical support and intelligence, Fertig's auxiliary force was one of the best in the Archipelago. The problem faced by Fertig and his commanders in providing food for their guerrillas was solved by cargadores. Cargadores (porters) brought food to his headquarters all the way from Misamis. The journey took one month and covered approximately 320 miles: 120 miles through Japanese controlled territory and another 200 over the mountains. The cargadores packed in rice, corn and camotes from Misamis, and to prevent malnutrition they ate poly vitamin pills brought by the submarines.

Fertig also had devised means for feeding his fleet of cargadores. To have food for travel, carabao meat was cut into thin strips, dipped in tuba vinegar and brine, and dried in the sun. The jerky-like substance, called tapa, cured in two days and would remain preserved indefinitely. Another common practice was to simply tie a live chicken to a waist belt.

The Mindanao guerrillas even had their own navy, consisting of a small coastal fleet which was a hodge-podge of boats. The captured *Nara Maru*, a 60-foot Japanese-made diesel motor launch, ran on coconut oil. The *Nara Maru* was armed with a .50 caliber gun salvaged from a smashed B-17. The gun had a recoil spring improvised from rubber tubing. In one unique case, a guerrilla "quartermaster" walked a carabao to the edge of a cliff, slit its throat, and rolled it over the edge of the cliff into the sea. It was taken aboard the *Athena*, the flagship of the guerrilla navy which was waiting below, cut up and cooked.[1] The *Athena* was a two-masted sailing ship mounted a muzzleloader fashioned from four-inch pipe which fired balls cast from melted fishing weights. She had a crew of 150 armed with 20 automatic rifles. Amongst the *Athena's* spoils of war was a Japanese Mitsubishi medium bomber. It was brought down with her 20-millimeter cannon, a submarine-type deck gun. Another unique ship was the *So What*, a 50-foot boat skippered by Waldo Neveling, a German citizen and soldier of fortune whom Fertig commissioned in the US Army. The *So What* was armored with steel circular saw, on her gunnels and was used to convoy supplies, to raid Japanese inter-island commerce and to protect the mouth of the Agusan River.

Another Mindanao guerrilla naval vessel was *The Bastard,* a 26-foot whaleboat which was captained by Australian Jock McLaren. She mounted a 20-millimeter cannon in the bow, two twin .30-inch guns amidships, and a .50 caliber gun slightly aft. She was unique because she mounted an 82-millimeter mortar in her stern. A feisty little ship, *The Bastard* unhesitatingly sailed into Japanese controlled ports in broad daylight, sprayed the wharf with automatic fire, shot mortars at Japanese boats, turned tail and ran. Her crew even challenged the Japanese to duels by sending them written invitations. She even stood-to and engaged enemy strafing aircraft as if she were a heavily armed battleship.

While ferrying supplies by "navy" was quick, supply efforts by foot and cart was slow. For example, while going from Iligan to Misamis took six hours by banca, it

---

[1] *They Fought Alone*, 314-315.

could take up to three days on foot for the 25-mile trek through the jungle. Slower still, the overland route from Misamis to the east coast of Mindanao took three weeks one way. To organize things, Fertig developed a courier system, much like the messenger systems of the earlier Greeks. The runners would travel unescorted except through areas patrolled by the Japanese. All of this was part of Fertig's intricate convoy system, which protected the inter-island delivery of supplies brought by submarine and transported inter-island by cargadores to all of Fertig's six divisions (See figure 12-1).

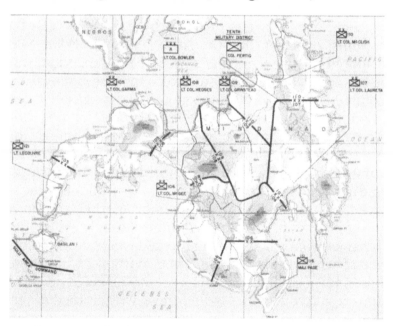

Figure 12-1. 10th Military District, Late 1944. Courtesy US National Archives.

News and Intel

The "bamboo telegraph" was faster but the message almost always became distorted in the sending. Guerrillas figured a 30-day period for Japanese Informants to travel the length of the Agusan River to Davao to report of guerrilla activities near Butuan to the Japanese commander at Army Headquarters. Even without cargadores a guerrilla could make only seven miles a day through jungle occupied by hostile pagans or Moros.[2]

News among the guerrillas could travel in another way and that was by newspaper. McCLish's 110th Division periodically published the *Free Man* with news received on shortwave radio broadcasts and in magazines brought by the submarines. Father Haggerty published the *Ateneo War News* prior to the surrender. Mindanao had fewer such "free press" papers than did some of the guerrilla organizations in the northern islands.[3]

In addition to looking to their spiritual welfare, the Church of Mindanao solved many of the guerrillas' communication and intelligence problems. The priests would perform their ecclesiastical duties during the day, and at night, many would dress like peasants and work with the guerrillas. They were the only truly secure means of sending messages between guerrilla commands, both on and off the island. There was no central clearing house to screen credentials of emissaries. To send an emissary without credentials "was to send him to sure death." Fertig used Father Hurley, a Jesuit Superior, to screen his emissaries, and Father Haggerty carried many messages within

---

[2] *Philippine Diary*, 250, 261.
[3] See Hawkins, *Never Say Die*, 167; *Guerrilla Padre*, 11.

Mindanao.[4] On Mindanao a priest was freed by the Church to serve his parish as his conscience dictated, and some even accompanied guerrillas on their combat missions.[5]

With great risk, civilians in many cases also worked as double agents and counterintelligence agents for the guerrillas. The Moros were especially good at this as that was their *modus operandi* when dealing with others in any case. Japanese counterintelligence was not very effective, for evidently their soldiers were not very circumspect in their private conversations with the Filipinos or among one another when Filipinos were near. For example, one Japanese intelligence report concludes that "Those [Filipino agents] who mingle among the men of the units quickly discover our activities and plans."[6]

In some instances, the propaganda war had a humorous twist. For example, the Japanese commander in Surigao published a leaflet offering a 1,000-peso reward for "the severed head of Sergeant Paul Marshall, US Army." Marshall made up a leaflet of his own offering a reward for the Japanese officer's head – then nailed it on the officer's door one night.[7]

Weapons and Maintenance

Production and maintenance of weapons on Mindanao was always a problem for the guerrillas. Fertig ordered his guerrillas to police up their cartridge

---

[4] Forbes J. Monaghan, *Under the Red Sun: A Letter from Manila*, 1946, 152.
[5] *They Fought Alone*, 133.
[6] Adalia Marquez, Blood on the Rising Sun, 1957, entire; *They Fought Alone*, 309.
[7] *Never Say Die*, 157.

casings after an ambush so they could be refilled and reused. Failure to bring back the casings could mean a demotion and loss of face. Fertig's guerrillas devised ingenious means. Bullets were made from lead poured into handmade sand molds or fashioned from ".30 caliber" curtain rods. Primers were made from mixing sulphur and coconut shell carbon. Some "small arms factories" could turn out as many as 160 bullets a day.[8] Amatol was removed from Japanese anti-ship mines and mixed with low-grade miner's dynamite to make powder for the cartridges. An alternative source was powder from duds and Chinese firecrackers. Fuses were made from tinfoil, potash permanganate and matchbox scrapings.

Cannons were made from brass pipes and catapults from bamboo and rubber inner tubing. Homemade grenades were made of coconuts charged with dynamite, or a dynamite stick with a short fuse would be placed in a tin can and the remaining space filled with nails, nine pieces of chain, nuts and bolts. The top of the can was sealed with pitch. Dynamite was found in many of the old mines on Mindanao. Incendiary bombs were made from beer bottles filled with gasoline, stoppered with a detonator and connected with a safety fuse – a rare device because there was rarely any gasoline.[9]

The guerrillas naturally preferred the .45 caliber Thompson sub machine gun and the .30 caliber Browning automatic rifle over their ancient weapons and homemade arsenal. But in a pinch, they would make a *paltik*, a homemade shotgun. With a block of wood, piece of water pipe, copper wire and a nail, a guerrilla

---

[8] Robert B. Osprey, *War in the Shadows: The Guerrilla in History*, Vol 1 (New York: Double Day, 1975), 515.
[9] Ibid, 179-180.

could make one of these devices designed to use a shotgun shell. They were said to be "effective." SWPA wanted to wean the guerrillas to the lightweight carbine, but in the meantime the Enfields, Springfields and captured Japanese rifles had to be repaired. An ejector spring for the Enfield could be fashioned in two days using only a hammer, chisel, rattail file and a steel strap from an automobile spring.[10]

Items other than weapons were also in short supply. Lye for soap was produced by first burning coconut palms and then making lime from roasting coral or sea shells. When blended together this created the lye. With coconut oil mixed with it the lye made a good lathering soap. It's one fault was that it tended to dye the hair a "brilliant henna." Another method of making soap was to boil shreds of coconut, then add an extract of hardwood ash to the coconut oil. The ash and oil when stirred and boiled together made soap. Ink for printing currency and making typewriter ribbons could be made by mixing soot with glycerin.

Perhaps the favorite necromancy was making the native tuba drink. To make tuba one first bled the sap from a frond on a coconut tree, then fermented it with pulverized tanbark from a mangrove tree. Alcohol was extracted by using a still made from a Socony can with bamboo tubes running beneath a stream for condensation. If the guerrilla lived in Cotabato Province he would extract alcohol from the mash of a gabi root (potato).

The alcohol was also used to run gasoline engines. But with a little egg, chocolate and sugar, the tuba made a potent 9.6 percent proof cocktail. Many of the guerrilla

---

[10] Ingham, *Rendezvous by Submarine*, 137-138; Keats, *They Fought Alone*, 152, 342.

units took great pride in bottling the best tuba in the area and sending the "best labels" to other guerrilla commanders as gifts. As McClish is purported to have mused on one occasion: "It's not high-test and it's not Old Grandad, but it'll get you there, or get you drunk."[11]

Airfields

On a scale which must have seemed like that used to build the Panama Canal, the Mindanao guerrillas built "Farm Project Number 2," a 7,000-foot runway in the middle of a giant rainforest. The labor crews took a year to build it. They worked at night by firelight and camouflaged the field by day. Fertig had begun building airstrips early because he thought they could be made useful to the guerrilla movement. He later built others when instructed to do so by GHQ to prepare the island for the American invasion. Mindanao was to be the anchor in the air assault on Formosa. The guerrillas built the airfields and then covered them over with topsoil and planted crops on them. All that was needed then was a bulldozer to scrape the dirt from the runway. Aircraft could land at night with torches burning at each end of the runway.

The Japanese were aware of the airfield construction, but seemed to make little effort to end the activity. In one intelligence report the Japanese concluded that the guerrillas had constructed a large underground hanger at the airfield near Domikan, a capability beyond even the ingenuity and determination of the Mindanao guerrillas.[12]

---

[11] Quote from Ingham, *Rendezvous Submarine*, 64, see also page 88; Ira Wolfert, *American Guerrilla in the Philippines*, 1945, 109.

[12] For information on guerrilla airfields see: U.S. Naval Institute, Reminiscences of Rear Admiral Arthur H. McCollum, USN, (Retired), 1973,

Radios

As to communications, the best that can be said about the radios used by the guerrillas is that they represented state of the art equipment. To carry the radio, its engine and its barrel of lubricating oil took a convoy of fifty cargadores. As an added difficulty, American radio equipment wasn't jungle proof, and was generally too large for loading into the submarines, and unthinkable to port through the mountainous rain forests of Mindanao. To reach Australia, Fertig used the American HT-9 transmitter rigged with a special parabolic-type antenna. But the HT-9 was American made and not able to withstand the wet tropical heat. It broke down often, and Fertig replaced it with a more rigorous Australian TW-12.

To reach the elements of his command, Fertig used a 3BZ radio transmitter which was also compatible with Australian radios at the time (ATR4As). ATR4A transceivers carried two-and one-half watts of power, and, on occasion, these radios would themselves reach Australia. To solve the problem of recharging, radio batteries were recharged by bicycles hitched to a generator. A more elaborate method used by Pendatun in Cotabato was to remove the differential and axles from a truck, bolt paddles to the wheel flanges, and suspend the device on a platform over a swift mountain stream. Power was transmitted through the differential into a drive shaft which was hooked up to a generator.

Tape 13, p. 596; Ltr, Fertig Lo Casey; Haggerty, Guerrilla Padre, pp. 193-197; Carlos P. Romulo, I Saw the Fall of the Philippines, 1942, p. 307; ATIS, I, pp. 3, 20, 24, 47, 71, 78; ATIS, II, pp, 55, 62.

The gears could be shifted to adjust to the rate of flow of the stream.[13]

Throughout Mindanao, Fertig maintained connectivity with coastwatcher stations. The coastwatchers, which consisted of any variety of folks, were successful in transmitting weather reports three times each day and reporting enemy ship and aircraft movements. Some of the results were spectacular. Guerrilla intelligence reports from Mindanao provided SWPA with information on Japanese naval movements, base locations, troop dispositions, air activity, and bomb damage assessments.[14]

Daily reports flowed from the clandestine radio network were the first thing that MacArthur read each day.[15] Some of these reports had significant effects. For example, guerrilla reports aided US Navy submarines in sinking over three hundred Japanese ships off the southern coast of Mindanao over two years.[16] In fact, on June 15, 1944, a coast watcher under Fertig's command reported Japanese ship movements through the San Bernardino Strait that gave early warning that allowed the US Navy to adjust its plan prior to the Battle of the Philippine Sea, aka the "Marianas Turkey Shoot."

---

[13] For information on radios see: Ind, *AIB*, 167, 207-209, 214; Edward Dissette and H. C. Adamson, Guerrilla Submarines, pp. 75-76; Ingham, *Rendezvous By Submarine*, 88, 157; Wolfert, American Guerrilla, p. 212; Kermit Roosevelt, War Report of the O.S.S. (Office of Strategic Services), 1976, pp. 137-1 40. The OSS had developed the SSTR series of radios – SSTR stood for Strategic Services Transmitter-Receiver -- and this is one case where the OSS would have been of great assistance to General MacArthur. The SSTR radios were used in the European Theater and elsewhere in the Far East, and they were much more sophisticated than those used by the guerillas.

[14] Holmes, *Wendell Fertig*, 136-43.

[15] *MacArthur's Undercover War*, 113-114.

[16] *Wendell Fertig*, 56.

The information from the coastwatchers was compared with that received from ULTRA - decoded Japanese electronic communications – to verify ship locations. Off the coast of Mindanao, a troop convoy of forty-nine merchant ships and light escort vessels moved just off Surigao enroute to Davao. The coastwatchers acted quickly and watched excitedly as Halsey's fleet units pinned the convoy to the shore. A Grumman pilot bombing Davao City had seen the convoy and had flashed the warning to the battle fleet. Thirty-two vessels were sent aground in the bays along the coast by American aircraft. The Filipinos appeared in large numbers with bolos to welcome the Japanese ashore to Mindanao. They left no survivors, and the ships were stripped of materials as they lay floundering on the coral reef.[17]

---

[17] See Monaghan, *Under the Red Sun*, 245; General Headquarters, Far East Command, Operations of the Allied Intelligence Bureau, 1948, 57; Dissette, *Guerrilla Submarines*, 112; Courtney Whitney, *MacArthur: His Rendezvous with History*, 135; Ind, AIB, 210.

Chapter Thirteen
War of the Flea

"Analogically, the guerrilla fights the war of the flea, and his military enemy suffers the dog's disadvantages: too much to defend; too small, ubiquitous, and agile an enemy to come to grips with. If the war continues long enough—this is the theory—the dog succumbs to exhaustion and anemia without ever having found anything on which to close its jaws or to rake with its claws." – Robert Taber

The guerrillas were in a dilemma when deciding how often to engage the Japanese. If they attacked too often, or were too successful, then the civilians were open to reprisals and the guerrilla bases became the objective of punitive expeditions. However, if they didn't attack enough, they could lose the confidence and support of the people. A factor that was to prove beneficial for Fertig and his commanders was that for the most part, the Japanese had been content to control only the larger towns and to leave the jungle to the guerrillas. With the exception of the concerted effort to eliminate Fertig and his guerrilla headquarters, the Japanese had not expended large numbers of men in pursuit of the guerrillas.

Then in late 1943 and early 1944 the Japanese launched some brutal operations throughout Mindanao in conjunction with their declaration to kill every living American still free in the islands. Fertig had gotten wind of these attacks from intelligence sources in Manila who said that General Jiro Harada, commander of the 100th Division, had been ordered to end the guerrilla resistance once and for all. The same sources said that an entire division of soldiers specially trained in anti-guerrilla tactics was being sent to Mindanao, but this was not the actual case. However, the Japanese attack did hit the island pretty hard. Hedges and Bowler barely escaped the net thrown over the island, and Fertig was driven deeper into the Agusan River Valley.

The Japanese were now preparing themselves for the coming invasion by American forces. By early 1944, it was clear what direction the war in the Pacific was taking. The Japanese high command estimated that throughout the Philippines a minimum of 24 battalions would be needed in the rear areas to guard against guerrillas with seven divisions needed to meet the invasion effort, a ratio of three front-line troops to every one soldier tied down in rear area security. The Japanese command was ultimately to conclude that "it is impossible to fight the enemy and at the same time suppress the activities of the guerrillas."[1]

The Japanese knew that MacArthur's instructions to the guerrillas was to organize, build strength and gather intelligence. And they understood the tactics being used by the guerrillas when they did engage the Japanese: "The enemy draws us out by using small units and then carries out an enveloping attack with his main force," or when outnumbered he lies in wait in the jungle "for our return and attacks fiercely."[2] But understanding the guerrilla tactics was not the same as defeating the guerrillas themselves. Ultimately, the Japanese were insulted and outraged at having to fight an enemy who would not give them a stand-up fight. Dealing with a foe who struck silently and quickly and then melted away, "like a fish in the sea," ran counter to their training and to their military code. This was ironic, of course, because

---

[1] Gene Z. Hanrahan, *Japanese Operations Against Guerrilla Forces*, 1954, 20.
[2] ATIS, II, p. 1; ATIS, I, p. 7.

it was the Japanese who were slaughtering innocent women and children.

The Japanese tactical method was to quickly arrive in force in an area with mortars, machine guns and plenty of ammunition and then deploy in the expectation that the guerrillas would accept the challenge to arms. "Their optimism was boundless, previous experience to the contrary, and they would plug away for four or five hours." The guerrillas would tease the Japanese with just enough fire to delay his advance, thereby permitting the civilians to escape into the jungle.

In the meantime, the guerrillas would withdraw to favorable ground in the hopes of luring the Japanese onto untenable ground. The Japanese were reduced to dumping leaflets from the air over guerrilla strongholds "calling the guerrillas yellow, urging them to come out and fight like men like the Japanese soldier, who is not afraid to die for his Emperor."[3]

Counterguerrilla Tactics

The Japanese employed much the same concept in tactics against the Filipino guerrillas as they did against the guerrillas in China. The Japanese had two types of operations: "alertness," which was conducting rear area security, and "mopping-up," which was an expansion of the geographical area to be occupied by using punitive expeditions and the standard tactical concepts of encirclement.[4] These tactics, regardless of how skillfully executed, were doomed to failure. Perhaps the biggest disconnect for the Japanese counterguerrilla operations

---

[3] *Rendezvous by Submarine*, 89, 107.
[4] Gene Z. Hanrahan, *Chinese Communist Guerrilla Tactics: A Source Book*, 1952, 132.

was they simply never recognized the political nature of guerrilla warfare. Principally this was because they never understood the nature of the people they were fighting. The counterguerrilla operations were, for the most part, conducted by occupation forces. In many instances the soldiers' first impulse was to solve the guerrilla problem by military means alone.

The few successes that the Japanese had were related more to the personality and character of an individual Japanese commander. This was the case in eastern Surigao Province where the Japanese commander had treated the people well and pacified the area, to Fertig's consternation.

There is no real evidence that the Japanese ever organized a counterguerrilla force or formulated a counterguerrilla strategy that was centrally directed. The better trained Japanese soldiers had been indoctrinated in the tactic of infiltration as a battle technique, and they had used it elsewhere when confronting large conventional forces. But they never adapted the technique to the counterguerrilla efforts and therefore failed to realize the full potential from their troops. When their counterguerrilla operations failed, the Japanese resorted to the one tactic with which they seemed comfortable, and one which required the least manpower and creativity from the commander. This tactic, of course, was terror.

To dissuade others from joining the guerrillas, the Japanese would hang the head of a local guerrilla at the entrance to the family's barrio with a sign on it reading, "bad man of the woods." Torture, internment and mass executions were commonly used tools, and entire areas were declared "bandit zones." The crops in these areas were destroyed, and the civilians ordered to leave. The

Japanese understanding of the impact that their terror tactics would have on people was good only to a point.

For example, guerrilla leaders were paranoid of strangers because the Japanese hired Filipino assassins to kill resistance leaders. This tactic had its intended effect because the Filipinos were unable to trust any but their closest friends, which is why the priests became so important for communicating personal messages between guerrilla commanders. But the Japanese failed to come to grips with the real effect of their policies which was that their policies actually drove more people to the guerrillas.

In some cases, the local Japanese commander simply made a deal with the guerrillas. For instance, when the Japanese commander in Misamis Oriental Province was unable to pacify the area, he negotiated an agreement with Governor Palaez. Palaez agreed to keep McClish's guerrilla force from attacking the garrisons if the Japanese would agree to stay out of the area. The Japanese commander demanded and received one concession, and that was the right to send patrols periodically through Medina, a town on Gingoog Bay. Palaez agreed, and every two months the people of Medina would leave the streets as the Japanese patrol passed through the town.[5]

In the last analysis, the Japanese tactics and policies failed to either pacify Mindanao or defeat the guerrillas. They blamed this failure first on not having enough troops to root out the guerrillas and second, the terrain of the island itself, which gave the guerrillas sanctuary. In addition, they lamented the "constant rampancy" of

---

[5] *Philippine Diary*, 254-255.

the Moro tribes and the guerrillas "radio activities" which brought the submarines to Mindanao.[6]

So much for the so-called Co-Prosperity Sphere. The Japanese were never able to accomplish what they came in the first place to do, that is, to exploit the wealth of the island: the lumber, chrome, iron, manganese and coal. The metrics for the Japanese weren't good: Two Japanese divisions plus their support troops were tied down on Mindanao, and these suffered continued attrition from guerrilla inflicted casualties and from disease. While it was beyond the capabilities of the guerrillas to bring the Japanese to any kind of decisive engagement, their tactics were, indeed, effective.

Overall, the effectiveness of the guerrillas can be measured by the continued resistance of the civilians to the Japanese and the necessity for the Japanese to station over two divisions of soldiers on Mindanao to combat the guerrillas. For their part, the Japanese continued throughout the occupation to suffer casualties, and the success of the guerrillas gave heart to the civilians so that they would continue to resist. With the Filipino people of Mindanao willing to provide critical logistics, intel and manpower, Fertig and his commanders were able to keep the home fires burning as the resistance to the Japanese become even more resolute.

Notwithstanding, by late August 1944, the situation on Mindanao and at Fertig's headquarters seemed grim indeed. By now, Fertig was at Waloe, which was deep in the bush. At this point, in order to communicate with Australia, Fertig had to employ heavy V-beam antennas. This made triangulation easier for the Japanese, and when the bombers arrived on target they could spot the

---

[6] GHQ, FEC, Japanese Monograph No. 3, 11.

antennas.[7] But just when things seemed to be at their darkest, the first American bombers to appear over the Philippines in two and a half years unloosed their bombs over Davao City. And from then on, one aircraft bombed Davao City each night. In all, twenty-two attacks were made from August 6-22, with 32 planes dropping 41 bombs. Then on September 1, 1944 Liberators and Lightnings dropped 130 tons of explosives on Davao. The effect was dramatic and immediate. The Japanese troops withdrew from the interior and hastened their preparations of the beach defenses.

Pillow Effect

Aside from MacArthur's orders to avoid combat with the Japanese, there were some imperatives of the guerrillas' circumstances which dictated that this would be so irrespective of MacArthur's orders. Although Fertig and his commanders desired to engage the Japanese in offensive operations, thereby violating the spirit of MacArthur's orders, pitched battles with the Japanese were not feasible because of the small amount of ammunition and number of weapons available. Furthermore, pitched battles meant heavier guerrilla casualties, and with no doctors or medicine with which to treat the wounded, most wounds were in the long-run mortal.[8] Fertig identified another factor when he wrote:

> They Filipino are damned fine guerrilla fighters, but they never will be first class combat troops, as we do not have officers to lead them, and absolutely no way of giving them the sound training necessary to make combat troops.[9]

---

[7] *They Fought Alone*, 316-317, 367-368.
[8] For example, see *Never Say Die*, 1961, 153.
[9] Letter Lt Col W. W. Fertig to General Hugh J. Casey of July 1, 1943.

So Fertig implemented what he called his "pillow defense."[10] The only ground the Japanese soldier occupied was that beneath his feet when he moved he gained new ground, but he lost all that he had. And as the Japanese would attack, the guerrillas would melt into the jungle and hillside. There would be nothing for them to concentrate their effort. With this tactic in mind, like Marion and Mosby, Fertig deployed the guerrillas near their own barrios. He believed that near their own homes they would fight better, and they would certainly be easier to feed and clothe. As such, the guerrillas rarely accepted combat with the Japanese near the barrios where they lived. When the Japanese patrols conducted a sweep through a barrio, the guerrillas would withdraw.

The Japanese feared the bruising guerrilla attacks when they did come, and at the Davao Penal Colony, where guerrillas would snatch soldiers guarding prisoners working in the fields, the Japanese would actually walk in the middle of a prisoner formation for security. Whereas elsewhere in the Pacific Theater the Japanese were feared for their night fighting abilities, in the Philippines the situation was quite the reverse. The Filipinos owned the night.

Moreover, the Filipinos would use *soyac* traps, pointed bamboo stakes driven in to one foot above the ground on both sides of a trail. As a Japanese patrol would come along the trail, the guerrillas would fire several shots and shout. The Japanese soldiers would dive into the high grass onto the spikes. The Filipinos would then kill them with bolos.[11]

---

[10] *They Fought Alone*, 225.
[11] Ingham, *Rendezvous By Submarine: The Story of Charles Parsons and the Guerrilla - Soldiers in the Philippines,* 1945, 108-109.

The guerrillas claimed that the Japanese soldier was easy to detect at night because he smelled bad from poor personal hygiene.[12] And during the daytime, the Japanese were easy to track because they wore distinctive "tabby toe" boots, a soft boot with a separate toe for the big toe. The Filipinos often took the uniforms from dead Japanese in order to have clothes to wear, and they looked much like Japanese soldiers because of their dark skins and small stature. But the Filipinos of necessity went barefoot and the Japanese never did, so that is how they could be distinguished from each other.[13]

Ambush

Japanese tactics were stereotyped, therefore predictable, and thus they were easily ambushed. They would move troops in "boxcars," huge trucks used to move sugar plantation laborers before the war. The boxcar could hold 150 troops. As one of the oldest forms of unconventional warfare, laying in an ambush requires some concealment, firepower, and a bit of intel as to which road or trail will be frequented and when. The Japanese always moved at the same hour of the day, and they rarely had a choice of roads on which to travel. This made ambushing them easy. And for their part, their tactics on contact were always the same, and ambushes became almost set-piece, choreographed affairs. After the guerrillas melted into the jungle, the Japanese would carry their dead tied on poles much like you would a wild

---

[12] *They Fought Alone*, 343.
[13] Ind, Allied Intelligence Bureau: Our Secret Weapon in the War Against Japan, 1958, 178.

pig carcass. They would even sometimes burn their dead or badly wounded on a funeral pyre at an ambush site.

The Japanese also related the increase in aggressiveness of the guerrilla to the increase in ammunition being brought in by submarines. They experienced disruption in their rear areas, especially in Bukidnon from Grinstead's forces. Moreover, Japanese intelligence on the size of the guerrilla force was never good. With the exception of the one time when they so badly overestimated Fertig's strength in Misamis City, they consistently underestimated the guerrilla numbers.

For example, Bowler believed that the Japanese estimated his strength at five percent of what it really was.[14] Overall, Japanese intelligence from early 1943 until 1945 put the 10th Military District at a nucleus of 100 Americans with 3,000 Filipino and American guerrillas. They never had the guerrilla strength above 6,000 on Mindanao, only one-sixth of the actual strength. However, they did have the order of battle and unit designations correct, except they called the 10th Military District "10 Army Group," using their own military notation, and they always had Fertig shown as "Major General" or on occasion as "Brigadier General."

It is difficult to conclude what size force the Mindanao guerrillas numbered at this time because even though there is an abundant quantity of personnel strength reports available, it must be assumed that the number of guerrillas was still fluctuating considerably as many recently freed from Japanese occupied areas hastened to join the guerrillas before the war was over. More than likely, by this time, the 10th Military District could boast a strength of 40,000 guerrillas.

---

[14] *Philippine Diary*, 259, 299.

Casualties for the guerrillas are harder to pin down. Robert Ross Smith concludes that there are no reliable casualty figures for the Philippine guerrillas accept for those on Northern Luzon.[15] The Japanese give their own figures, plus estimated guerrilla casualties in their intelligence reports. As a sample, the Japanese reported their own casualties as 13 officers and 325 enlisted killed in action and 20 officers and 454 enlisted wounded for all of the Philippines during a five-month period January through May 1944.

For the month of June 1944 alone, the Japanese claimed that the Kyo Group on Mindanao engaged 2,690 guerrillas in 51 engagements, capturing 79 and killing 75. The Kyo Group was one of four groups.[16] This report showed 18 Japanese killed in June on all of Mindanao. The Mindanao guerrillas claimed 100 Japanese casualties for every one of their own. Father Haggerty concluded that over the period of the Japanese occupation he conducted 300 times as many baptisms as he did funerals, and most of the funerals were conducted for people who had died of malaria.[17]

As to enemy metrics, by this time, Mindanao guerrillas had killed some 7,000 Japanese, while tying up some 60,000 Japanese troops intent on breaking the resistance.[18]

---

[15] Robert Smith, *Triumph in the Philippines*, 692.

[16] ATIS, I, p. C; ATIS II, 13-14; Figures for Mindanao alone can be reconstructed using the 10-day reports of each punitive expedition.

[17] Haggerty, Guerrilla Padre in Mindanao, 1946, 66.

[18] By October 1, 1944, the Japanese complement of 13 divisions and five brigades deployed in the entire Philippine Archipelago numbered some 288,028.

Chapter Fourteen
The Jungle Telegraph

*"Guerrilla action reverses the normal practice of warfare,
strategically by seeking to avoid battle and tactically by
evading any engagement where it is likely to suffer losses."*
– B. H. Liddell Hart

The most important mission that SWPA had given to
the Philippine guerrillas was that of gathering
intelligence. In a report made to the US Congress after
the war, the information gathering effort by the Filipinos
received this assessment: "One of the most dramatic
examples of practical intelligence in the war, in the
Southwest Pacific Area, is represented in the
development of the Philippine underground."[1] Carlos
Romulo, had this to say:

> Japanese plans, copies of their most secret advices,
> military dispatches, accounts of troop movements,
> number and location of enemy planes, all had been
> reported by native patriots directly to GHQ. The entire
> Japanese plan in the Philippines lay open here for
> General MacArthur to see and set his plans by...
> Everything was carefully worked out between a
> powerful Force working on the outside and a weaker
> but no less valiant force working from within.[2]

The relationship between SWPA and the Philippine
resistance movement was a classic demonstration of
how Otto Heilbrunn describes the cooperation between
a regular force and its client guerrilla force. The ideal

---

[1] Quoted in Charles Andrew Willoughby, *The Guerrilla Resistance
Movement in the Philippines: 1941-1945*, 1972.
[2] Carbos P. Romulo, *I See the Philippines Rise*, 1946, 37-38.

relationship has two primary facets: the partisans collect and pass on information to the sponsoring army, and the army "seconds its own intelligence officers to the partisans."[3]

For their part, the Mindanao guerrillas were also involved with relaying information from other islands as well as sending on their own. There were the usual sources of information available to the guerrillas, such as civil servants and service industry people. There were also some other, less usual, sources. Priests were generally treated as neutrals, so they were able to move freely in Japanese occupied areas and thus could carry intelligence as well as personal messages to the guerrilla leaders. In Davao, Illocano natives were hired by the Japanese to work as laborers in ammunition dumps, at airfields and in the Japanese headquarters in Davao. Unbeknownst to the Japanese, many of these young men, though thought to be ignorant, had college educations and bilingual capabilities.

However, Japanese sources were not so productive, for Colonel Kobayashi, the 14th Area Army operations officer, claimed that "While the Americans steadily received intelligence from their guerrillas, our group never gave us any information that we could use."[4] To ensure that their information was useful, the guerrillas often went to great lengths. For example, when reporting the location, number and type of antiaircraft guns, the guerrillas would sometimes make pencil-and-paper rubbings of the guns' serial numbers on the

---

[3] Otto Heilbrunn, *Partisan Warfare*, 1962, 113.

[4] Supreme Commander for the Allied Powers, Reports of General MacArthur: The Campaigns of MacArthur in the Pacific, Volume 1, 1966, pg. 302.

identification plates to prove the accuracy of their information.[5]

The consensus, was that the Mindanao guerrillas were proficient in the gathering of information for use by the invading forces for planning the Victor V operation. The name given for the upcoming US invasion of Mindanao. In preparation for this, the commanding general of Eighth Army, General Robert L. Eichelberger had directed his subordinate commanders to utilize the guerrillas for the gathering of information. His guidance was as follows: Commanders will "utilize guerrilla forces for information gathering agencies and establish direct signal communication between the local guerrilla intelligence net and his own headquarters."[6]

Users of the information received from guerrillas on Mindanao agreed that the sketches of enemy positions and concentrations and hand-drawn maps showing details not shown on photographs were invaluable. Maps were especially important because even though there had been extensive surveying by Army engineers of the area before the war, the printing of the maps had not been completed before the Japanese attacked. In fact, the guerrillas on Mindanao and in the Visayas used oil company maps and individual sketch maps.[7]

The estimates of enemy strength was another matter. As all sources generally agree, the estimates of Japanese troop strength was invariably high. Some sources say that estimates were "exaggerated," but that implies a willful manipulation, which may or may not have been the case.[8] Willoughby diplomatically finds in this

---

[5] *They Fought Alone*, 402.
[6] Headquarters Eighth Army, Field Order No. 26, March 20, 1945.
[7] Wendell W. Fertig, "Guerrillero," Part I, no date, 54.
[8] Headquarters, Eighth Army, Report of the Commanding General Eighth Army on the Mindanao Operation, no date, 104; 31st Division, Historical

propensity to overestimate the enemy strength a "richness and variety" in guerrilla reporting. However, MacArthur's historians wrote that "within its limits of accuracy" the information from Mindanao was very useful in planning invasion operations. Parroting this sentiment, the 31st Infantry Division called their intelligence reports "models of accuracy," but tempered this praise by stating that their real value was realized when the reports were carefully collated with other sources. Willoughby agrees that the use of aerial photos combined with the guerrilla reports gave a highly reliable picture of the actual situation.[9]

In the last analysis, the information provided by Fertig's guerrillas was so accurate that the Eighth Army found it gained little by sending in their own intelligence agents ahead of the coming invasion. Moreover, if they were to be detected, they could compromise the secrecy of the landing. As such, Eichelberger prohibited any teams from entering Mindanao before the landings there.[10]

For obvious reasons, the most accurate picture of Japanese troop strength on Mindanao was a necessity. While Eighth Army estimated enemy strength to be 34,000, Fertig surmised the estimate of Japanese strength at 69,140.[11] As for the Japanese, they were convinced the guerrilla intelligence was accurate, for they were on the receiving end when the bombs struck

---

Report, pp. 71-73; Department of the Army, Office of the Chief of Military History, "The Philippines Guerrilla Resistance Movement," no date, pp. 226-227 (hereinafter OCMH, "Guerrilla Resistance Movement"); Commander Amphibious Group Si::, U.S. Pacific Fleet, "Report of Amphibious Attack on Zamboanga, Mindanao," March 26, 1945, pp. 52-53.
[9] Charles A. Willoughby and John Chamberlain, *MacArthur 1941-1951*, 1954, 215; SCAP, Reports, 1, 309; 31st Division, Historical Report, 66.
[10] HQ, 8th Army, Report: V-III, V-IV.
[11] Tenth Military District, Intelligence Summary, No. 13, Feb 1945, 6.

home. Colonel Ind relates the story that the Japanese released an official communique which declared the Americans had "perfected a new aerial bomb which was attracted by concentrations of ammunition and fuel."[12] However, the guerrilla intelligence effort was not without its detractors. While he officially encouraged the use of guerrilla intelligence, Eichelberger had nothing good to say in his unofficial comments. He went so far as to say, "no credence is to be given guerrilla reports and tactical decisions are not to be affected by them."[13] However negative were these views, Eichelberger did write of Mindanao that "we did have considerable information about dispositions of enemy troops. Since guerrilla forces on Mindanao were the most efficient and best organized in the Philippines."[14] However, he also said: "Part of my personal aggressive policy in Mindanao...was based on erroneous intelligence of the Japanese strength."

Eichelberger would later push the 24th Division hard, and had them strung out over 50 miles, but the problem was not so much overestimation of the Japanese strength as it was underestimation of their strength. For example, Eichelberger was also to claim that the guerrillas overestimated the Japanese strength as they followed the Japanese retreat up the Agusan River Valley.

The 24th Infantry Division observed that "information of the energy and of the Mindanao roads and trails prior to the operations was sketchy." Moreover, guerrilla reports greatly exaggerated the enemy strengths and dispositions. Civilian reports were much more reliable.

---

[12] Ind, AIB, 156.
[13] Luvaas, Dear Miss Em, 251.
[14] Eichelberger, Jungle Road, 217.

Suffice it say, the Mindanao guerrilla intelligence reports were not always accurate; as the 162nd Infantry found when it followed a guerrilla sketch map and attacked in the wrong direction from the enemy's position.[15] Certainly, it is difficult to estimate enemy strength in a jungle, and Fertig believed that the increased movement of Japanese units because of air attacks made the reporting even less reliable.

To be balanced, the 31st Infantry Division had this to say: "Guerrilla reports were usually more accurate than civilian reports reflecting, naturally, at least a bit of military training."[16] So it seems that, just as with any military operation anywhere in the world, there was both good and bad intelligence information supplied to the invasion forces. But Fertig had a solution, albeit macabre, to the problem of accuracy and credibility. Keats elaborates:

> Once, when Headquarters disbelieved Fertig's estimate of Japanese casualties, Fertig sent them two demijohns filled with matched pairs of ears that the Moros had collected. Headquarters never publicly doubted Fertig's estimates of enemy casualties thereafter.[17]

---

[15] William F. McCartney, *The Jungleers: A History of the 41st Infantry Division,* 1948, 148.
[16] 31st Division, Historical Report, p. 73.
[17] *They Fought Alone,* 411.

Chapter Fifteen
VICTOR V

"Guerrilla warfare is the culmination and the active expression of a successful resistance movement." – Russell Volckmann

The war in the Pacific was now in its fourth calendar year and the embattled Filipinos and their guerrillas were eagerly awaiting a US invasion. By this point, their Japanese captors had inflicted a heavy toll of atrocities. In the emotional high of the much-anticipated US invasion, the guerrillas began to strike openly against Japanese forces. In preparation, Fertig also directed his forces to provide every available bit of intelligence on the enemy forces, beaches, roads, bridges, and airfields, etc.

Early on, the Joint Chiefs of Staff had concluded that Mindanao would be the anchor for the assault on "the vital Luzon-Formosa-China Coast Area." Mindanao was selected because of Fertig's USFIP had several thousand guerrillas under his command and were operating on and controlling most of the island. It was believed that Mindanao had the most to offer an invasion force. The plan was for Mindanao to be reduced and secured in two months.[1]

The coming invasion had been carefully nurtured. Submarines and aircraft had supplied Fertig's command, which was now 33,000 on the rolls, with arms, ammunition and supplies of all sorts. In preparation, Fertig had organized his USFIP into six

---

[1] Joint Chiefs of Staff, Future Operations in the Pacific: Report by the Joint Staff Planners, March 10, 1944, pp. 10, 21; See also General Headquarters, Southwest Pacific Area, "Basic Plan for Montclair Operations," February 25, 1945, pp. 3-4 for the concept.

divisions, augmented by the Maranao Militia Force and a loosely organized "division of Moros.[2]

Code-maned Operation Montclair, the US invasion of Mindanao was set for November 15, 1944 with a subsequent strike at Leyte Gulf on December 20, 1944. Mindanao was to be designated GOA #1 (Guerrilla Operational Area Number One) of 14 GOA's. But Admiral Halsey discovered a weakness in the air defense of the Philippine Islands over Mindanao while supporting landing operations in Morotai with his carrier task forces. His pilots tested the air defenses over the Visayas and found them to be weak as well. He recommended an immediate change in the initial landing site, and MacArthur agreed.

The Montclair plans were scrapped, and plans for Operation Victor were quickly drawn up. Leyte would now become the site of the first assault on October 20, 1944. Although the guerrillas on Mindanao did not know it at the time, Mindanao, the second largest island in the archipelago, would be the 21st island to be invaded.[3]

Back on 25 December, 1943, Yamashita had directed Lt. Gen. Sosaku Suzuki, 35th Army commander, to evacuate his forces from Leyte as best he could and make preparations to defend the rest of the southern and central Philippines, including Mindanao. Yamashita, of course, had long since written off the 100,000-man 35th Army. Nonetheless, Suzuki was to tie down as many Allied divisions as possible in the Visayas and Mindanao. To that end, Suzuki planned to make his stand in east-central Mindanao where he "hoped to set up a little self-

---

[2] Robert Smith, *Triumph in the Philippines* (Washington, DC: Center of Military History, United States Army, 1963), 586.
[3] *MacArthur: Rendezvous*, 152.

sustaining empire that could hold out indefinitely."[4] However, due to Allied naval interceptions, Suzuki failed to evacuate the whole of his army from Leyte, leaving 20,000 behind. Overall, as Robert Ross Smith observes: "by February 1945, the time was long past when the Japanese on the southern islands could hope for anything more than to die while conducting a static defense."[5]

Under Suzuki's command was the 30th and 100th Divisions with dispositions as follows: Lieutenant General Gyosaku Morozumi commanded the 30th Division and had overall command of all Japanese forces east of Lake Lanao, while Lieutenant General Harada commanded the 100th Division and forces in the rest of the island. These were not first-rate troops. In fact, of the 55,000 strong Japanese force on Mindanao, only some 15,000 of which could be considered anything approaching combat effective.

The 30th Division, the better of the two, had come from Korea but had lost four of its nine battalions to Leyte where they had been annihilated. The eight battalions of the 100th Division had been living the easy life. Not more than ten of its officers were regulars, and the quality of the junior officers and noncommissioned officers was "lamentable." A third of the division were Korean conscripts, soldiers not normally enthusiastic for their fate decreed by Emperor Hirohito. The troops were poorly trained and the units widely scattered. They were understrength, poorly equipped and their communications were inadequate. The leaders had a defeatist attitude, and the troops were complacent because they had been by-passed by MacArthur's forces.

---

[4] Smith, *Triumph in the Philippines*, 587.
[5] Ibid.

Nevertheless, they felt that they could cope almost indefinitely with the guerrillas, a less worthy foe in their estimation than the American forces. But guerrilla attacks and air strikes had destroyed most of their transportation capability, and they had just enough military supplies to defend initially against a conventional invasion force, although not enough to sustain the fight. If left to fight only the guerrillas they would fare better. Fertig's guerrilla force of 36,000 would have had a tough time dislodging such a force which could still be formidable.

Having experienced great disappointment when the first blow had not fallen on Mindanao, the guerrillas of Mindanao certainly didn't relish the prospect of engaging single-handedly the by-passed Japanese troops. Their angst was assuaged on 10 March 1945, when MacArthur directed Eighth Army under Eichelberger to sweep the Imperial Japanese 35th Army off of Mindanao.[6]

Code-named Operation VICTOR V, the planned campaign would take four months. Major units available to Eichelberger's Eighth Army included X Corps headquarters, the Americal Division, the 24th, 31st, 40th, and 41st Infantry Divisions, and the separate 503d Parachute RCT. All told, Eichelberger's force numbered

---

[6] D. Clayton James, *The Years of MacArthur, Volume II, 1941-1945,* 1975, 104.
After the landing on Luzon, MacArthur never really had authority to continue the liberation of the southern islands in the archipelago. Had Eighth Army troops and shipping been needed elsewhere at the time, it is doubtful if these additional landings would have been conducted. At Yalta in February 1945 the American Joint Chiefs had told their British counterparts that they had no intention of sending major American forces to conquer the southern islands. General Marshall "assumed that the Filipino guerrillas and the newly activated Army of the Philippine Commonwealth could take care of the rest of the country."

42,000 combat troops and some 11,000 service troops (See figure 15-1).

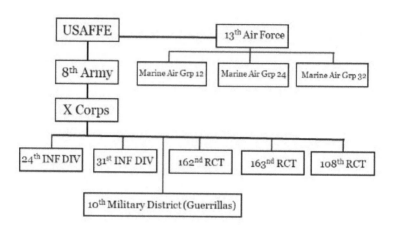

Figure 15-1. 8th Army organizational Chart.

The success of the guerrillas in fighting along-side the regular forces was much the same that they had in providing intelligence information: there were mixed reviews. There had to be cooperation between the invasion forces and the guerrillas so that the operational plans of the invading forces were not adversely affected by guerrilla activities. Coordination was needed to ensure that bridges, roads or facilities required by the attacking forces were not destroyed. The commander had to ensure that guerrillas did not affect enemy movements planned for by the conventional force, such as attracting undesired reinforcements or preventing the movement of reserves in response to a feint.[7]

In the broad concept the guerrillas would launch an offensive prior to the American assault to clear or isolate

---

[7] For example, see Heilbrunn, *Partisan Warfare*, 113-114.

objectives. Then they would fight along with the conventional units to secure further objectives, and, finally, the guerrillas would conduct the mopping-up operations. In order to satisfy the command and control over the guerrillas necessary to carry out these tasks, the commander X Corps was given operational control over guerrilla units attached to his combat units. The Commander, 10th Military District retained administrative control.[8]

To maximize potential, Eighth Army had a Guerrilla Subsection in the general staff G-2. The expectation was that the guerrillas would be used to the maximum in sabotage and harassing operations and intelligence gathering.[9] Eighth Army gave many missions to the guerrillas which they were expected to fulfill on Mindanao. These missions were very diverse, and for the most part the guerrillas carried out the missions as expected. They would initially assist advance parties by providing security and information to signal and intelligence teams and hydrographic survey parties. They would provide combat intelligence, guides, interpreters, and reinforce communications and reconnaissance agencies.

Moreover, they could assist pilots downed behind Japanese lines and conduct harassing ambushes and sabotage in enemy rear areas. Guerrillas could destroy aircraft and coastal guns prior to the invasion as well as attack areas deep in the enemy's rear after the invasion. They could provide labor for local working parties, cargadores for supply movements, guards for prisoners

---

[8] Headquarters, X Corps, Field Order No. 38 "Operations on Mindanao," 30 June 1945, p. 2.
[9] GHQ, SWPA, Staff Study, V-4, p. 2; Smith, Triumph in the Philippines, p. 586; General Headquarters, Southwest Pacific Area, Operations Instructions No 91, February 14, 1945, 11.

of war and provide security guards for roads, supply dumps and key bridges. The guerrillas could provide supplemental supply in certain rare instances, assist with the evacuation of the wounded, and provide a military police function to restore order in liberated areas. The guerrillas could assist civil affairs units in identifying and interrogating collaborators and in working with the local civil governments. Finally, although their capabilities were limited to do so, the guerrillas could fight alongside the conventional force as a conventional unit integrated into the tactical planning and organization. They would more usefully be employed in mopping-up operations in this regard.[10]

In preparation for the coming invasion, Fertig's guerrilla units were given some pre-assault missions such as cutting Japanese overland lines of communications, clearing prospective beachhead areas, and attempting to bottle Japanese forces into small areas.[11] The 10th Military District now had the green light to "go big" on the Japanese, and they soon "added greatly to the woes of Morozumi and Harada," with the demolition of supplies, roadblocks, bridge destruction and a myriad of attacks on Japanese garrisons.

---

[10] The sources which contain samples of these various capabilities of the Mindanao guerrillas are too numerous to list here. Histories, staff studies, personal accounts and field orders already cited contain examples. Four that contain some interesting examples but which have not been cited above are: Headquarters, USAFFE, "Operation of the Counter Intelligence Corps Detachment, 41st Infantry Division at Zamboanga, Mindanao, Philippine Islands," USAFFE Board Report No.256, April 16, 1945; History of the 31st Infantry Division Training and Combat, 1940-1945, 1946; Headquarters, Third Engineer Special Brigade, Eighth Army, Third Engineer Special Brigade Historical Report, 1 May 1945 to 31 May 19451, June 11, 1945; Charles W. Boggs, Jr., Marine Aviation in the Philippines, 1951.

[11] *Triumph in the Philippines*, 586.

As to intel, guerrilla reports pinpointed targets for bombing, thereby eliminating the need for an advanced aerial reconnaissance. Aircraft were destroyed on the ground by sabotage. The guerrillas had solutions for all sorts of problems. Eighth fighter strip was too far away to support the coming invasion. This problem was solved in a somewhat novel manner. Fertig's 105th Division, under Lt. Colonel Hipolito Garma "had long held a good, prewar landing strip at Dipolog, on the north coast of the Zamboanga Peninsula."[12]

In fact, during January and February 1945, the guerrillas had seized the Dipolog airstrip and had held it while surrounded by Japanese. On March 8, 1945 two reinforced companies of the 21st Infantry, 24th Division flew aboard C-47s into Dipolog to reinforce the defense of the airfield. That same day sixteen Marine Corsairs arrived. The airfield's close proximity would later enable Marine pilots to fly out of Dipolog and carry out bombing missions in Zamboanga City 150 miles away.

The invasion of Mindanao began with a naval and air bombardment along the beach defenses east of Zamboanga City. This forcing the IJA 54th Independent Mixed Brigade (IMB), under Lt. Gen. Tokichi Hojo, to abandon its excellent defensive positions along the southern shore of the Zamboanga Peninsula. Leaving behind a few scattered outposts, Hojo withdrew his force 8,000-man force to higher ground a few miles inland.[13] As the 162nd and 163rd Infantry Regiments (of the 81st Division) made their landings, Hojo's force fled into the hills in disorder. The coastal plain was cleared of enemy troops by dusk on 11 March.

---

[12] Ibid., 591-592.
[13] Ibid., 593.

As the 162d and 163d Infantry Regiments moved northward from the coast, Eighth Army directed Fertig to block the escape route of Hojo's 54th IMB. Fertig entrusted this task to Capt. Donald J. LeCouvre's 121st Infantry, 105th Division, to block the east coast road at Bolong. This forced Hojo's remaining 5,000 troops into the interior.[14] Guerrillas added further to the advance of Eighth Army by assuming garrison duties, particular on the Island of Basilan, relieving the 162nd.

Since early March, Lt. Colonel Hedges' 108th Division had moved against the Japanese garrison at Malabang. Receiving some close air support from Marine Corsairs based out of Dipolog, Hedges was able to make such headway that Allied planes could now use the Malabang airstrip. This development made sweeping changes to X Corps' plan. By 5 April Marine aircraft started operating from the field, and by the 11th of April the last Japanese troops fled toward Parang, giving the guerrillas complete command of the entire Malabang region. On 13 April, Fertig radioed Eighth Army that X Corps could land unopposed at Malabang and Parang.

The landing site was changed to Parang at the last moment, and the Malabang airfield was used by Marine Air Group 24 (MAG-24) to support the drive eastward across Mindanao. Then on April 17, 1945 the 24th and 31st Infantry Divisions of X Corps landed at Cotabato, drove east to Davao City and north to Join the 108th Regimental Combat Team which landed at Macajalar Bay on May 10th (See figure 15-2). This invasion strategy nearly duplicated the Japanese attack on the American-held island in 1942.

---

[14] It is estimated that some 6,800 men of Hojo's 54th IMB died in the retreat.

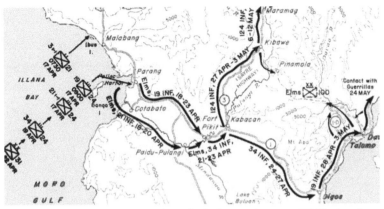

Figure 15-2. Main US Invasion of Mindanao. Courtesy US National
Archives.

The guerrillas also supported the advancing forces as
they pushed the Japanese farther back into the interior.
Guerrilla units were attached to regular units and shifted
from one to the other. There were some unconventional
units formed, even in comparison to the already unusual
organization of regular and guerrilla units. Lieutenant
Colonel Bowler led an attack on Japanese positions at
Sarangani Bay with an oddly configured provisional
infantry battalion. The battalion was made up of
antiaircraft troops from Battery B of the 496th
Antiaircraft Gun Battalion, acting as infantry, and a
guerrilla combat company from the 118th Infantry,
106th Division. The battalion used engineer LCM's and
was supported by Marine close air support. Also
participating in this attack were elements of the guerrilla
11th Infantry and the 108th Division. The expeditionary
battalion received very high marks for its fighting
qualities.[15]

---

[15] *Triumph in the Philippines*, 647.

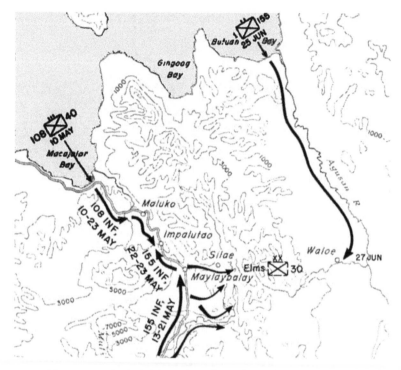

Figure 15-3. Main US Invasion of Mindanao (Macajalar Bay).
Courtesy US National Archives.

Generally speaking, however, the use of the guerrillas in the US troops in some cases were actually unaware that Filipino soldiers were also fighting the Japanese in their area, and this made integrated tactical formations especially difficult and hazardous. The Commanding General, Eighth Army concluded that the guerrillas performed their many other missions well but "it is a mistake to use them in the attack as they are critically short of equipment and have little understanding of the tactical principles involved in offensive combat."[16]

---

[16] HQ, 8th Army, Report: V-Ill, V-IV, p. 95; see also Richardson, One-Man War, p. 151.

On June 30, 1945 General Eichelberger announced that the organized enemy resistance on Mindanao had ceased with the capture of Davao City, and the victory operation was officially declared closed with mopping-up and security missions continuing. Actually, the fight to clear Mindanao of organized bodies of Japanese continued after that date. In fact, soldiers of the 24th Division regarded the post-Davao operations as "the hardest, bitterest, most exhausting battle of their ten island campaigns of the war."[17] However, there can be no doubt by 30 June, the main goals of the campaign had been realized.

By August 15th, American Army casualties had reached 820 killed and 2,880 wounded. Of the 55,850 Japanese in Central Mindanao, 47,615 were accounted for by August 15th as dead, wounded or surrendered. That left 8,285 Japanese – using Japanese sources – for which there was no accounting. These soldiers probably slipped into the jungle and died of starvation, disease, or fell prey to the guerrillas, or Moros.[18] Figures are nowhere recorded for estimated guerrilla casualties.

With Japanese forces broken and retreating into the Waloe area, which had been the last location of Fertig's headquarters before the American invasion, X Corps assigned the primary operational mission to the 10th Military District to "establish and maintain contact with hostile forces."[19] With Frank McGee commanding, the 107th Division took over from the 24th Division in late July and continued to hunt down Japanese stragglers until the surrender on August 15th. McGee was felled by

---

[17] Ibid.

[18] *Triumph in the Philippines*, 647-648, 692, 694.

[19] Headquarters, X Corps, Field Order No. 37 (Operations in Mindanao), June 19, 1945.

a sniper's bullet on August 7th, signaling symbolically the end of the American-led guerrilla movement on Mindanao.

Upon liberation, the Philippine Islands reverted to its pre-war US commonwealth status. Then on July 4th, 1946, as promised, the Commonwealth of the Philippines became the Republic of the Philippines. Manuel Roxas was its first president.

Chapter Sixteen
Conclusion

"It is not too late to learn from the experience of history. In any case, those who frame policy and apply it need a better understanding of the subject (guerrilla warfare) than has been shown in the past" – B.H. Liddell Hart

Shortly after his memorable return to Leyte, in glowing terms, General MacArthur summed up the enormous contribution the guerrillas had made to the present US invasion with the following words:

As our forces of liberation roll forward the splendid aid we are receiving from guerrilla units throughout the immediate objective area and adjacent islands causes me at this time to pay public tribute to those great patriots both Filipino and American who had led and supported the resistance movement in the Philippines since the dark days of 1942. These inadequately armed patriots have fought the enemy for more than two years. Most are Filipinos but among these are a number of Americans who never surrendered, who escaped from prison camps, or who were sent in to carry out specific missions.

Following the disaster which, in the face of overwhelming superior enemy power, overtook our gallant forces, a deep and impenetrable silence engulfed the Philippines. Through that silence no news concerning the fate of the Filipino people reached the outside world until broken by a weak signal from a radio set on the Island of Panay which was picked up, in the late fall of that same fateful year, by listening posts of the War Department and flashed to my Headquarters. That signal, weak and short as it was, lifted the curtain of silence and uncertainty and disclosed the start of a human drama with few parallels in military history.

In it I recognized the spontaneous movement of the Filipino people to resist the shackles with which the enemy sought to bind them both physically and spiritually. I saw a people in one of the most tragic hours of human history, bereft of all reason for hope and without material support, endeavoring, despite the stern realities confronting them, to hold aloft the flaming torch of liberty. I gave this movement all spiritual and material support that my limited resources would permit.

Through the understanding assistance of our Navy I was able to send in by submarine, in driblets at first, arms, ammunition and medical supplies. News of the first such shipment spread rapidly throughout the Philippines to electrify the people into full returning consciousness that Americans had neither abandoned them nor forgotten them.

Since then, as resources increased, I was enabled, after formalizing the guerrilla forces by their recognition and incorporation as units of our Army, to send vitally needed supplies in ever increasing quantities through Philippine coastal contacts by four submarines finally committed exclusively to that purpose. I would that at this time I might name the gallant heroes of this epic in Philippine-American history, but considerations of security for the individuals, their families and the cause require that I limit myself to a generalization of their work and a statement of their brilliant achievements.

Of the latter I need but point out that for the purposes of this campaign we are materially aided by strong, battle tested forces in nearly every Philippine community, alerted to strike violent blows against the enemy's rear as our lines of battle move forward and that now are providing countless large areas adjacent to military objectives into which our airmen may drop with assurance of immediate rescue and protection. We are aided by the militant loyalty of a whole people-a people who have rallied as one behind the standards of those stalwart patriots who, reduced to wretched material conditions yet sustained by an unconquerable spirit, have formed an invincible center to a resolute over-all resistance.

We are aided by the fact that for many months our plans of campaign have benefited from the hazardous labor of a vast network of agents numbering into the hundreds of thousands providing precise, accurate and detailed information on major enemy moves and installations throughout the Philippine Archipelago. We are aided by the fact that through a vast network of radio positions extending into every center of enemy activity and concentration throughout the islands, I have been kept in immediate and constant communication with such widespread sources of information.

We are aided by the fact that on every major island of the Philippines there are one or more completely equipped and staffed weather observatories which flash to my Headquarters full weather data morning, afternoon and night of every day and which in turn provides the basis for reliable weather forecasts to facilitate and secure the implementation of our operational plans. Widely disseminated to our forces throughout the Pacific and in China the information from this weather system has materially aided our military operations over a large section of the world's surface.

We are aided by an air warning system affording visual observation of the air over nearly every square foot of Philippine soil established for the purpose of flashing immediate warning of enemy aircraft movement through that same vast network of radio communications. We are aided by provision of all inland waterways and coastal areas of complete observation over enemy naval movement to give immediate target information to our submarines on patrol in or near Philippine waters. This information has contributed to the sinking of enemy shipping of enormous tonnage, and through such same facilities was flashed the warning to our naval forces of the enemy naval concentration off the western Philippines during the Marianas operation.

Finally, we are aided by the dose interior vigilance that has secured for our military use countless enemy documents of great value, among which were the secret defensive plans and instructions of the Commander-in-Chief of the combined Japanese areas and complete information on the strength and dispositions of enemy fleet and naval air units. That same Commander-in-Chief of the Combined Japanese Fleets was a prisoner of one of our guerrilla units prior to his death from injuries sustained in an air crash.

All of these vital aids to our military operations, and there are many more still unmentioned, are responsive to the indomitable courage of the military and civil leaders whom I shall in future name and their loyal followers both Filipino and American; to gallant Filipinos, residents of the United States, who have volunteered to infiltrate into the islands in succor of their countrymen and Americans who have infiltrated with them; and finally to the militant loyalty and unconquerable spirit of the masses of the Filipino people.

As Commander-in-Chief of the forces of liberation I publicly acknowledge and pay tribute to the great spiritual power that has made possible these notable and glorious achievements-achievements which find few counterparts in military history. Those great patriots, Filipino and American, both living and dead, upon whose valiant shoulders has rested the leadership and responsibility for the indomitable movement in the past critical period, shall, when their identities can be known, find a lasting place on the scroll of heroes of both nations-heroes who have selflessly and defiantly subordinated all to the cause of human liberty. Their names and their deeds shall ever be enshrined in the hearts of our two peoples in whose darkest hours they have waged relentless war against the forces of evil that sought, through ruthless brutality, the enslavement of the Filipino people.

To those great patriots to whom I now pay public tribute I say stand to your battle stations and relax not your vigilance until our forces shall have swept forward to relieve you.

There are many lessons which can be drawn from a study of the resistance movement on Mindanao. The trap which awaits the unwary in arriving at these lessons is the likelihood that the rhetoric of the heroic literature on the subject will seduce the reader into advocacy and sympathy rather than objective appraisal of the information. This was a heroic people who waged a desperate fight against a cruel conqueror. An estimated 1,000,000 Filipinos died in the war, and this with no Philippine battle fleet at sea or grand army in the field.

It is very difficult to read the personal accounts of life under the occupation and not feel personally involved somehow. Still, a dogged effort will reveal certain truths which consistently appear throughout the reading, and it is from these facts, distilled and closely scrutinized, that the picture of the Mindanao resistance movement emerges.

A resistance movement requires leadership to provide organization, vision, direction, and coordination. In light of the loathed Japanese policies, there were no real ties to break between the Filipinos and the government of occupation. However hated were the Japanese, Fertig still needed to create an area command on Mindanao which would link popular support to the guerrillas and in turn reciprocate the guerrillas' support back onto the population. Fertig's initial approach of organizing the resistance and giving it a vision and direction was not enough to draw enough popular support. And so, like most other successful resistance movements, outside support in the form of both material and moral was needed. While General MacArthur actualized this support in material form to sustain the fight, moral support to sustain the will to fight came from President Quezon which in turn brought the hearts and minds of the people. With this material and moral sanctioning in place, and an understanding of Filipino demography, cultures, taboos and beliefs, Fertig and his commanders were able to achieve unity of command over the guerrilla forces on Mindanao. This unity of command, combined with material and moral support became the center of gravity for the resistance movement on Mindanao. All of this constitutes points 1, 2, 5 and 6 from our introductory reasons for success.

Additionally, the terrain on Mindanao favored the guerrilla, and its obstacles could be overcome to some degree by modern communications – the radio. Guerrilla groups could be widely separated within secure mountain sanctuaries yet still be bonded together by communications. A knowledge of how to survive in the jungle and mountainous terrain was perhaps the Mindanao guerrillas' greatest asset (point 4).

One aspect of the resistance seemed counterintuitive. The culture of the Filipinos on Mindanao would seem to have militated against the successful establishment of an organized resistance movement because of the language, religious, and social differences among the populace. But here is where the American leaders played a crucial role in bringing unity to the movement. The Americans were considered neutral in the struggle for political power on the island, and ultimately the Americans were the only antagonists against the Japanese who had nothing to gain from leadership within the movement other than survival and revenge upon the Japanese. But if the Americans were the thread which drew the diverse Filipino groups together, the Japanese themselves provided the mortar which held them tightly bound.

As to the external support, the material provided by SWPA undoubtedly contributed to the overall success of the resistance movement far beyond its apparent capacity to do so. The ammunition and weapons helped the guerrillas sustain the fight at a minimum level, and the radios helped bind the guerrilla groups together. But the real significance of the submarine visits was two-fold: through the recognition of leaders and the shipment of supplies to them, General MacArthur was able to tie the guerrilla force together under one central command and give them credibility in the eyes of their fellow Filipinos.

Moreover, the submarines brought hope to the Filipino people. The submarines represented the keeping of a promise, and they were a tangible sign that one day the Filipino people would be rid of the conqueror.

There are many other lessons which can be drawn from the resistance movement on Mindanao, but there are an equal number which do not have clear answers. The success of the Mindanao guerrilla movement would seem to say something about the type of leaders best fitted to lead a guerrilla resistance, and it does to a point. But the leaders on Mindanao were unique men who in many ways were ideally suited to lead the movement.

They were at home in the culture and environment to large degree. They had reputations for being businesslike and apolitical. They were older and presumably wiser. More important, most of them possessed *bona fide* military credentials – United States Army commissions – which gave them widely accepted credibility. Their success was greater because they were on Mindanao than it might have been elsewhere because of the unusual situation caused by the need to have neutral leaders for the Moros and Christians.

The contribution of the guerrillas to the success of the American invasion is clear, and it is widely accepted that many lives of American soldiers were saved through the contributions of the guerrillas. Still, just how much they contributed to the success of the invasion depends on the objective sought to be achieved. Did they fight as regiments alongside American regiments? No. Did they save lives and hasten the invasion of the Philippines?

Yes. The question which this suggests to us today is: Do we fully appreciate the full potential from a guerrilla force, and by extension, do our military and State Department leaders today fully understand how to do so?

This question provides a lead to further research that might be pursued. In addition, it is tempting to propose research which will compare the resistance movement in the Philippines with the resistance movements elsewhere during World War II. When it is eluded to at all, the Philippine resistance is often passed aside as an "internecine struggle." But I suspect that the Philippine resistance was far more widespread and united within the population then in many other resistance movements and as well led as most.

The most fertile area open to further research into the Mindanao resistance movement itself is an oral history which could be conducted with the surviving American members of the Mindanao guerrillas. A major contribution could be made on how the Mindanao guerrillas were first organized and what personal relationships – politics, decisions – affected the growth of the organization. In many respects this book has only barely tapped the potential for the study of this guerrilla organization.

Perhaps more than any other single factor, the occupation policies of the Japanese did more to unite the Filipino people in resistance. For many Filipinos, the issue was clearly one of survival. For others the broader, more abstract ideological issues of freedom and democracy drove them into resistance. For some, the test became simply one of good versus evil. Both the Christians and the Moros saw the Japanese occupation as a threat to their religion and an assault upon their personal system of values.

A proud people, the Filipinos on Mindanao did what they had done historically – they resisted. And for whatever reasons, the Japanese government never understood that the basic tenets of their occupation policies could have only the one predictable effect: to compel the Filipino people to resist.

It may be said that the swiftness of Japanese victories in the Pacific left them wholly unprepared to cope with the administration of the conquered countries. They lacked the trained colonial administrators and technicians to run the banks, mines and factories. With no colonial civil service, the inexperienced bureaucrats assigned for service in the Philippines were full of condescension toward the local people and were largely ignorant of their customs and traditions. In fact, they found it easier to bully and intimidate than to understand the differences of culture. They turned to the military and the military answered to no one.

As previously said, though the Japanese maintained as much as 150,000 troops on Mindanao, they held back from committing them to counterguerrilla actions.[1] They simply failed to understand that they were fighting more than just isolated groups of bandits. Moreover, the terrorism and brutality practiced by the Japanese was predestined to be counterproductive, a fact well-demonstrated in the history of resistance movements. Fertig assessed the situation in Mindanao in this way:

> It was the civilians whom the Japanese butchered, and as long as the Japanese were bestial, popular support

---

[1] *War in the Shadows*, 518.

of the resistance was assured. And as long as there was resistance, the Japanese would continue to be bestial.[2]

Almost certainly, had the Japanese policies been genuinely benevolent, the resistance movement would have been far less successful. But the almost incredible stupidity of the Japanese occupation policies demonstrated that the Japanese conquerors had routinely underestimated or ignored the role their own policies played in producing the conditions from which a resistance movement could grow.

Again, turning to Sun Tzu, we find that the Japanese ignored the ancient wisdom available to them: "Those who excel in war first cultivate their own humanity and justice and maintain their laws and institutions. By these means they make their governments invincible."[3]

In light of this, as well as the above reasons, the Mindanao resistance movement remains one of the most successful on record. For the student, the Mindanao resistance movement provides a model of a successful resistance and provides some clues to the requirements for success of any resistance movement. It is the best, most complete example of unconventional warfare that is easily referenced in US history.

For the man-at-arms, it is an example of courage, ingenuity and duty. For the strategist, the Filipino resistance is an example of a successful resistance and should be used for studying United States' policies for acting in this arena.

As for Fertig, he is one of three men who used their wartime experience to formulate the doctrine of unconventional warfare that became the cornerstone of

---

[2] *They Fought Alone*, 196.
[3] Sun Tzu, *The Art of War*, 88.

Special Forces. Along with Aaron Bank, and Russell Volckmann, he is considered one of the founding fathers of the US Army Special Forces.

Appendix A
10th Military Division Units

The following notations outline the structure of the 10th Military District as of January 31, 1945:

10th Military District Headquarters

Established: September 18, 1942
Commanded by: Colonel Wendell W. Fertig
Personnel: Chief of Staff - LtCol Sam Wilson (Manila businessman); Deputy Chief of Staff - Maj. M. M. Wheeler (USNR); G-2 - Maj. H. A. Rosenquist; Signal Officer - Capt. James Garland

"A" Corps, Western Mindanao, Headquarters

Established: January 1, 1944
Commanded by: LtCol Robert V. Bowler (38 years old, taught economics at Washington State U. Owned fisheries in Alaska. Reserve officer called to active duty before the war).
Personnel Strength: officers - 142; enlisted - 798.
Personnel: G-2 - Maj. Chandler B. Thomas; G-3 - Capt. Donald H. Wills.

Units: 105th, 106th, 108th, 109th Divisions, 121st Separate Regiment and the 116th Separate Battalion.

105th Division

Established: January 23, 1943
Commanded by: LtCol Hipolito Garman (Filipino)
Personnel: officers - 324; enlisted - 4,270.
Units: 106th Regiment, 107th Regiment, 115th Regiment, 121st Separate Regiment - Lt. Donald Lecouvre (former U.S. Air Corps enlisted man). This regiment fell also under "A" Corps.

106th Division

Established: October 7, 1943
Commanded by: LtCol Frank McGee (Mindanao planter before the war. Former U.S. Army officer with World War I service. West Point graduate).
Personnel: officers - 298; enlisted - 3,595.

Units- 116th Regiment - Maj. Herbert Page (68 years old, former Philippine Constabulary officer. Mindanao planter at Glan, Cotabato before the war), 118th Regiment - Maj. Salipada Pendatun (Filipino), 119th Regiment, 116th Separate Battalion. Operated independently under "A" Corps also.

107th Division

Established: May 1, 1944
Commanded by: First commander was LtCol Clyde C. Childress (Battalion commander in 61st Division, PA, before the war). Childress was evacuated to join US Amy forces on Leyte January 1945. Succeeded by LtCol Claro Laureta, a Filipino PA officer.
Personnel: officers - 141; enlisted - 2,308.

Units: 130th Regiment, 111th Provisional Battalion - Lt. Owen P. Wilson (unsurrendered sergeant, U.S. Air Corps), 112th Provisional Battalion - Lt. Anton Haratik, Sternberg Detachment - Lt. Adolph Sternberg, Jr. (unsurrendered sergeant, U.S. Air Corps).

108th Division

Established: December 14, 1942
Commanded by: LtCol Charles W. Hedges (Age - late forties. With Kolambugan Lumber Mills on Mindanao before the war. Held U.S. Army Reserve commission. After Japanese invasion he commanded Gen. Fort's Motor Transport Co.)
Personnel: officers - 974; enlisted - 13,012.

Units: 105th Regiment; 108th Regiment; Maranao Militia Force - separate command (for political reasons), 124th Regiment, 126th Regiment, 127th Regiment, 128th Regiment, 1st Provisional Regiment, 2nd Provisional Regiment, Separate battalions - 4 units, Separate companies - 5 units.

109th Division

Established: March 14, 1943
Commanded by: Originally commanded by LtCol Robert V. Bowler. Succeeded by LtCol James R. Grinstead (age in mid-fifties. Retired U.S. Army officer - service in Mexico; World War I: twice wounded, received DSC; Philippine Constabulary officer in Moro campaigns. Mindanao planter in Cotabato before the war. Also succeeded to command 106th Division).
Personnel: officers - 327; enlisted - 3,987.

Units: 109th Regiment, 111th Regiment, 112th Regiment - Capt. William McLaughlin (former sergeant, 31st Infantry; commissioned when war broke out), 117th Regiment.

110th Division

Established: September 15, 1942
Commanded by: LtCol Ernest E. McClish (Manila businessman before the war. Nov. 1941 assumed command of a PA regiment being mobilized. Left hospital rather than surrender to Japanese. Evacuated to Leyte January 1944).
Personnel: officers - 317; enlisted - 5,086.

Units: 110th Regiment, 113th Regiment, 114th Regiment - Capt. Paul H. Marshall (U.S. Army Private First Class who escaped from Davao Penal Colony April 3, 1943 with 9 other Americans. Succeeded McClish on his evacuation to Leyte).

Notes:

(1) There were changes from time-to-time in the guerrilla unit leadership. This listing reflects only the January 31, 1945 leadership as shown in General Headquarters, US Army Forces, Pacific, "The Guerrilla Resistance Movement in the Philippines," March 20, 1948.

(2) Some 187 Americans have been listed as having fought with the guerrillas in Mindanao. Not all can be named here for the positions all held are not in the available records.

(3) After "A" Corps was established, Fertig maintained the 107th and 110th Divisions under his immediate control, while Bowler, with all of the remaining units, reported to Fertig through "A" Corps Headquarters.

(4) The figure below shows the operational boundaries of these guerrilla divisions in the 10th Military District.

Mindanao Guerrilla Organization
January 1945

110th Div

105th Div

109th Div

108th Div

107th Div

121st Rgt

107th Div

116th Bn

Bibliography

Abaya, Hernando J. *Betrayal in the Philippines*. New York: A.A. Wyn, 1946.

Agoncillo, Teodoro A. *The Fateful Years: Japan's Adventure in the Philippines 1941-45, Volume 2*. Manila: University of the Philippines Press, 1965.

Asprey, Robert B. *War in the Shadows: The Guerrilla in History*. Vol. 1. 2 vols. Garden City, NY: Doubleday, 1975.

Bernstein, David. *Philippine Story*. New York: Farrar, Straus and Company, 1947.

Breuer, William B. *MacArthur's Undercover War: Spies, Saboteurs, Guerillas, and Secret Missions*. Edison, NJ: Castle Books, 1995.

Guardia, Mike. *American Guerilla: The Forgotten Heroics of Russell W. Volckmann*. Philadelphia: Casemate Publishers, 2010.

_____. *Shadow Commander: The Epic Story of Donald D. Blackburn*. Philadelphia: Casemate Publishers, 2011.

Haggerty, Edward. *Guerrilla Padre in Mindanao*. NY: Longmans, Green, and Company, 1946.

Hart, B. H. Liddell. *Strategy*. New York: Meridian, 1991.

Hawkins, Jack. *Never Say Die*. New York: Dorrance, 1961.

Holmes, Kent. *Wendell Fertig and His Guerrilla Forces in the Philippines: Fighting the Japanese Occupation, 1942-1945*. Jefferson, NC: McFarland, 2010.

Hogan, David W. Jr. *Special Operations in the Pacific*. Washington, D.C.: Department of the Army Center for Military History publication, 1992.

Bibliography

Hunt, Ray C. and Bernard Norling. *Behind Japanese Lines: An American Guerilla in the Philippines*. Lexington: The University Press of Kentucky, 1986.

Hurley, Vic. *Swish of the Kris: The Story of the Moros*. New York: E.P. Dutton, 1936.

Keats, John. *They Fought Alone: A True Story of a Modern American Hero*. New York: Turner, 2015.

Kenworthy, Aubry S. *The Tiger of Malaya: The Story of General Tomoyuki Yamashita and "Death March" General Masaharu Homma*. New York: Exposition Press, 1953.

Krueger, Walter. *From Nippon to Down Under: The Story of the Sixth Army in World War II*. Washington, DC: Combat Forces, 1953.

Lapham, Robert and Bernand Norling. *Lapham's Raiders: Guerillas in the Philippines 1942-1945*. Lexington: University Press of Kentucky, 1996.

Lukas, John D. *Escape from Davao: The Forgotten Story of the Most Daring Prison Break of the Pacific War*. New York: Simon & Schuster, 2010.

MacArthur, Douglas. *Reminiscences*. New York: McGraw-Hill Book Co., 1964.

McCollum, Arthur H. *Reminiscences of Rear Admiral Arthur H. McCollum, US Navy (Retired)*. Annapolis, MD: United States Naval Institute, 1971.

Miksche, Ferdinand. *Secret Forces: The Technique of Underground Movements*. New York: Praeger, 1950.

Ney, Virgil. *Notes on Guerrilla Warfare: Principles and Practices*. Washington, D.C.: Command Publications, 1961.

Norling, Bernard. *The Intrepid Guerillas of North Luzon*. Lexington: University Press of Kentucky, 1999.

Ramsey, Edwin P. and Stephen J. Rivele. *Lieutenant Ramsey's War: From Horse Soldier to Guerilla Commander.* New York: Knightsbridge Publishing Co., 1990.

Richardson, Hal. *One-Man War: The Jock McLaren Story, 1957.* Sydney: Angus & Robertson, 1957.

Romulo, Carlos P. *I Saw the Fall of the Philippines.* New York: Doubleday, Doran & Company, 1942.

Smith, Robert R. *Triumph in the Philippines.* Washington, DC: Center of Military History, United States Army, 1963.

_____. *United States Army in World War II, The War in the Pacific, The Approach to the Philippines.* Washington, D.C.: Department of the Army, 1953.

Smythe, Donald. *Guerrilla Warrior: The Early Life of John J. Pershing.* New York: Scribner, 1973.

Steinberg, Rafael. *Return to the Philippines.* New York: Time Life Books, 1979.

Toland, John. *The Rising Sun: The Decline and Fall of the Japanese Empire, 1936-1945.* New York: The Modern Library, 1970.

United States Army Field Manual 3.0, Operations. Washington, D.C.: Headquarters, Department of the Army, October 2008.

United States Army Field Manual 3-05, Army Special Operations Forces. Washington, D.C.: Headquarters, Department of the Army, 20 September 2006.

United States Army Field Manual 3-05.130, Army Special Operations Forces Unconventional Warfare. Washington, D.C.: Headquarters, Department of the Army, 30 September 2008.

United States Army and Marine Corps Field Manual 3-24: Counterinsurgency. Washington, D.C.: Headquarters, Department of the Army, December 2006.

Bibliography

Volckmann, R. W. *We Remained: Three Years Behind the Enemy Lines in the Philippines.* New York: W. W. Norton & Company, Inc., 1954.

Von Clausewitz, Carl. *On War, Everyman's Library* (Princeton, NJ: Princeton University Press, 1993.

Whitney, Courtney. *MacArthur: His Rendezvous with History.* NY: Praeger, 1977.

Willoughby, Charles A. *The Guerilla Movement in the Philippines.* Vol. I. Intelligence Series. General Headquarters: United States Army Forces, Pacific, 1 March 1948.

_____. *Intelligence Activities in the Philippines During the Japanese Occupation, Documentary Appendices.* Vol. II. Intelligence Series. General Headquarters: United States Army Forces, Pacific, 10 June 1948.

Willoughby, Charles A. and John Chamberlain. *MacArthur: 1941-1945.* New York: McGraw-Hill Book Co, 1954.

Wolfert, Ira. *American Guerrilla in the Philippines.* NY: Simon and Schuster, 1945.

Zedong, Mao. *On Guerrilla Warfare.* Trans., Samuel B. Griffith, II. Chicago, IL: University of Illinois Press, 1961.

Index

Allied Intelligence Bureau (AIB), 65, 68

Abaca, 12

Amok, 22, 57

Anting-anting, 22, 59

Banca, 22, 84, 95, 102, 127, 131, 135

Barrio, 22, 33, 34, 36, 100, 106

Bowler, W., 24, 48, 49, 89, 146, 154, 171, 186

Carabao, 22, 31, 40, 62, 70, 75, 91, 127, 134

Cargadore, 22, 75, 127, 131, 134, 136, 142, 167

Childress, C., 48, 187

Clausewitz, Carl von, ix, 1, 3, 4

Counterinsurgency (COIN), xiv, 6, 7, 37, 104, 105, 108, 110

Datu, 22, 55, 57, 60, 91, 98, 115

Eichelberger, Robert L., 158, 160, 165, 173

Fertig, Wendell W., xi, 4, 11, 23, 28, 38-52, 58-63, 70-82, 88, 95-98, 145, 150, 162, 181

Fort, Guy O., 19, 20, 38, 39, 60

Garman, H., xxiv, 186

Grinstead, James R., xxiv, 48, 154, 188

Haggerty, E., 21, 35, 95, 137, 155

Halsey, William F., 144, 163

Harada, J., xxv, 145, 164, 168

Hedges, C., xxiv, 38, 39, 41, 48, 61, 85, 89, 99, 146, 170, 187

Homma, M., xxv, 17, 18, 20, 113, 114, 191

Juramentado, xxii, 56

Kalibapi, 6, 22, 104, 110

Katipunan, 8, 9

Kempeitai, 101-106, 113, 121

Kris, xxii, 54, 58

Laurel, Jose P., xxiv, 108, 109, 112

Laureta, Claro G., xxiv, 187

Lawrence, Thomas E., 4

Lecouvre, Donald J., xxiv, 170, 186

MacArthur, D., xxiv, 5, 6, 10, 11, 18, 22, 29, 44, 59, 64-69 78, 81, 97, 108, 112, 116, 118, 126, 143, 151, 163, 165, 176, 182

Magic Eye, 106

Maintenance, 138

Makeshift Ammunition, 139

Marshall, George C., 18, 138

Mao Tse-Ting,

McClish, Ernest E., xxiv, 48, 49, 70, 89, 97, 102, 137, 141, 149, 188

McGhee, F., xxiv

Medicine, 32, 86, 100, 115, 119, 151

Mestizo, xxii, 42, 75

Military Scrip, 72

Misamis City, 50, 84, 85, 87, 95, 98, 109, 154

Morgan, L., 42-46, 49, 62, 84, 96, 98-99

Morimoto, I., xxv, 50, 87, 94, 98, 114

Moros, 23, 44, 52-62, 99, 137, 161, 163

Morozumi, G., xxv, 164, 168

Office of Strategic Services (OSS), 66-67, 132

Operation Victor V, 162

Osmeña, S., 27, 74, 76

Page, Herbert C., xxiv, 48, 187

Parsons, Charles "Chick" T., xxiv, 23, 69, 85, 89, 91, 96, 117, 121, 131

Pendatun, S., 89-92, 99, 122, 187

Peralta, M., 68, 81-82, 121, 129

Pershing, J., 54, 55, 58

Prisoners of War (POWs), 18, 36, 39, 93, 95, 106, 107, 109, 132, 133, 152, 167

Propaganda, 33, 65, 66, 110, 122, 138

Quezon, Manuel L., xxiv, 6, 10, 11, 18, 27, 72, 176

Roxas, M., xxv, 174

Seven Dynamics of a Resistance, 5, 6

Seven Phases of a US Sponsored Insurgency, xvi

Sharp, William F., 18, 20, 21, 23, 38

Spyron, 69, 117, 124

Smith, C., 78, 79, 121

Suzuki, S., xxv, 163, 164

Tanaka, Y., xxv, 50

Unconventional Warfare (UW), Xiv, 3, 5, 6, 64-69

Wainwright, Jonathan M., xix, 18, 20, 22, 23

Whitney, C., xxiv, 65, 68

Willoughby, C., xxiv, 38, 66, 158, 159

Wilson, S., 48, 71-72

Yamashita, T., xxv, 163

Zonification, 106

About the Author

Colonel (Ret.) Larry S. Schmidt retired from the Marine Corps in 1994 after a career that spanned twenty-six years and numerous conflicts. Beginning with Vietnam, in 1968-69, he served as a Rifle Platoon Commander with the Leathernecks of B Co, 1st Bn, 5th Marines, operating in I Corps. From his experiences fighting the Viet Cong, he gained firsthand knowledge of the dynamics of insurgency. Following his service in Vietnam, he distinguished himself in various command and staff assignments which culminated with Operation Desert Storm (1990-91), where he commanded the 8th Marines during the liberation of Kuwait.

Colonel Larry S. Schmidt, Commanding Officer 8th Marines, Operation Desert Storm 1990-91.

Notes

Notes

Notes